ANIMAL SCIENCE, ISSUES AND PROFESSIONS

DEER

HABITAT, BEHAVIOR AND CONSERVATION

ANIMAL SCIENCE, ISSUES AND PROFESSIONS

Additional books in this series can be found on Nova's website under the Series tab.

Additional E-books in this series can be found on Nova's website under the E-book tab.

ANIMAL SCIENCE, ISSUES AND PROFESSIONS

DEER

HABITAT, BEHAVIOR AND CONSERVATION

A. A. CAHLER
AND
J. P. MARSTEN
EDITORS

Nova Science Publishers, Inc.
New York

For permission to use material from this book please contact us:
Telephone 631-231-7269; Fax 631-231-8175
Web Site: http://www.novapublishers.com

Library of Congress Cataloging-in-Publication Data

Deer : habitat, behavior and conservation / editors, A.A. Cahler and J.P. Marsten.
p. cm.
Includes index.
ISBN 978-1-62100-676-3 (hardcover)
1. Deer--Ecology. 2. Deer--Effect of human beings on. I. Cahler, A. A. II. Marsten, J. P.
QL737.U55D3967 2011
599.65'5--dc23
2011037410

Published by Nova Science Publishers, Inc. † New York

CONTENTS

PREFACE

This book presents topical research in the study of the habitat, behavior and conservation of deer. Topics discussed include the influence of anthropogenic disturbances on activity, behavior and heart rate of roe deer; deer as biomonitors of radioactive contamination; hunted red deer management; genetic distinctness of two endemic subspecies of deer in Korea and the effects of sika deer on vegetation in Japan.

Chapter 1 – Human recreation influences deer in various ways. To better understand these interactions three roe deer and two red deer were investigated during controlled disturbance trials. By means of implanted telemetry transmitters heart rate and positions of the head were recorded. Heart rates and behavior were compared before, while and after confronting the animals with various stimuli. Furthermore heart rate and activity of 14 roe deer and 8 red deer were continually recorded to obtain the normal daily and yearly patterns without disturbance trials. Altogether 6.8 million minutes of data were analysed. For each registered minute a distinction between active and inactive behavior could be made.

Daily activity patterns of roe deer varied greatly within different seasons: From November to June the animals were more active in the daytime, whereas in late summer and early autumn they showed pronounced nighttime activity. In contrast, red deer showed a relative constant ratio of day to night activity. Both roe and red deer were most active during twilight, roe deer more so at dawn, red deer more at dusk. Resting heart rate of red deer was lowest in winter (~44 beats/min), and highest in May and June (~71 beats/min). Roe deer showed less pronounced seasonal differences of resting heart rate (65-76 beats/min). Lunar cycles influenced activity of red deer, but not of roe deer.

Both roe deer and red deer exhibited stronger behavioral reactions ("activation") and heart rate increases towards optical stimuli (walking persons, dogs, horse riders and hang-gliders) than to acoustic stimuli (chain saw, radio music, dog barking) with exception of gunshots, which also elicited high reactions. Activation was associated with daytime and season. Both roe and red deer were more often activated in the morning hours and late afternoon than at noontime. Altogether the reactions of roe deer towards stimuli were short and strong, whereas red deer showed less obvious but persisting reactions. Behavior of red deer was disturbed for at least 10 minutes after gunshots and persons passing by, while horse riders, hang-gliders, tractors and most acoustic stimuli had milder reactions as consequence.

Roe and red deer showed different strategies of compensating after being disturbed. By roe deer the length of resting bouts were shortened on days with disturbance trials, but the number of resting bouts per day increased, so that the total activity budgets for daytime as well as for 24 hours were not influenced by disturbance trials. Red deer were more active during daytime when disturbed without changing their night activity (total daily activity increased), however, they had lower resting heart rates on days with disturbance trials. Recommendations are given to minimize disturbances of deer.

Chapter 2 – Adequate radiological protection of the environment requires the identification of animals sensitive to potential variations of the radionuclide content in their surroundings. Deer are considered suitable candidates for biomonitors in forest ecosystems. Due to their wide geographical distribution and their varied diet, the ICRP (International Commission on Radiological Protection) considers deer a reference animal. Deer also constitute a significant pathway for the transfer of radionuclides to man, especially in areas where a major deposition of anthropogenic radionuclides, such as $^{134,137}Cs$ and ^{90}Sr, occurred. The uptake of radionuclides takes place via the deer's diet. The radionuclide content in the plants they eat depends on various factors: plant species, soil properties, and the amount of the deposition, among others. The consumption of mushrooms is also an important source of radionuclides in deer. Depending on their chemical properties, these radionuclides are accumulated in different tissues of the animal's body. Radiocæsium is a chemical analogue of potassium, and is located preferentially in muscle tissue, and hence is important in the deer–man pathway. Also, naturally occurring radionuclides of the uranium series, such as ^{210}Pb and ^{210}Po, accumulate in muscle. Radiostrontium has a different chemical behaviour, as it is a chemical analogue of calcium. The ^{90}Sr is mainly detected in calcified tissues, such as bones and antlers. Because the antlers are shed annually, they constitute an excellent biomonitor for the evaluation of the temporal and spatial variations of radiostrontium in the environment. However, some factors, such as the age of the animal or its health status, may affect calcium remobilization, and hence cause certain variations in the radiostrontium content of the antler.

Chapter 3 – Many wild deer populations in the world are considered as threatened and deserving protection measures for population increase; in contrast, some managed deer populations are presently increasing to the point of becoming a nuisance or a permanent pest. In France, free ranging *Cervus elaphus* populations recently showed a strong increase, the global annual hunting plan rising from 10,000 to 71,000 individuals over the last 30 years. Frequently the local picture was an alternation of population peaks followed by crashes, with an approximate 10-years cycle. Such an evolution results largely from the ineffectiveness of the exhaustive counting methods previously used for deer management. Considering both the importance of estimating vital rates for managing deer populations and the extreme difficulty of obtaining reliable data in the field, the authors devised a new method to survey populations, in order to provide managers with a simple and effective tool. Their method takes advantage of the rich data sample provided by the annual cull, and the fact that hinds are shot randomly in driven hunts; hence the hind cull is a reliable representation of the hind population age structure. The mandibles of all animals killed by hunters are collected and analyzed, which allows the calculation of the ratio of yearling hinds in the total hind cull. Measuring mandible lengths is also an indicator of the populations physical condition. The authors conducted a survey of the deer population of Paimpont forest (Brittany, France) over ten years, which showed that the standard management methods underestimated population increases, while their yearling hind ratio allowed to more efficiently diagnose population increase and predict population crash. Their method has also been continuously applied since 2007 for deer management in the Indre District (annual cull of some 2000 deer) with promising results. The yearling hind ratio in culls is a simple management tool, dispensing with difficult field counts and allowing for gradual annual correction of management plans. Their study also demonstrated that this approach is applicable in this country on the scale of a whole district, given ad hoc imposed legislation and active contribution of hunting associations and management authorities.

Chapter 4 – Bullets used for deer shooting have traditionally and in many areas continue to, contain lead. Lead is a known toxin and poisoning can occur via several routes, including ingestion. Lead poisoning of waterfowl has led to regulations and agreements limiting the use of lead shot over wetland areas in many countries. More recent evidence has shown further significant issues in terrestrial habitats and affecting a wider range of species, notably birds of prey and other scavengers via secondary poisoning. These problems also extend to fragments of lead bullets. The standard butchering process often does not remove these ammunition fragments (from both shot and bullets) and contamination of the human food chain resulting from this can be sufficient to lead to increased risk of some health problems, particularly in high risk groups. Efficacious alternatives to lead bullets and shot have been produced and are readily available in most areas. Both voluntary and regulatory approaches have been attempted to reduce exposure to lead from ammunition sources, but these have so far been inadequate with reference to terrestrial areas and lead bullet use. Voluntary and regulatory approaches each have their costs and benefits and the most appropriate course of action will likely vary between situations and locations. While it is difficult to highlight the most appropriate response as a result, there is little doubt that further response is required by stakeholders, including Governments, if the problem is to be effectively addressed.

Lead has long been commonly used in ammunition and continues to be the base substance for most bullets used in a variety of situations worldwide. Its high density and relatively cheap price lent itself to production of accurate ammunition types able to transmit sufficient power to achieve rapid death of the target. For this reason, lead has traditionally been the most widely used base substance in manufacture of most ammunition types, including shot gun pellets, air gun pellets and rifle bullets.

Despite this wide use, lead is also known as a significant toxin to many living organisms. A wide range of toxic effects of lead are known. Injurious effects of lead are numerous and can be caused by sustained low levels of exposure. For example, in humans, there is evidence of permanent adverse effects of lead on cognitive function in children with blood lead levels below 10 $\mu g\ dL^{-1}$ (Canfield *et al.* 2003; Lanphear *et al.* 2005; Chandramouli *et al.* 2009). A tolerable weekly intake of 25 μg/kg b.w. is currently recognised, although the European Food Safety Authority's Panel on Contaminant in the Food Chain has concluded that this is no longer appropriate, and declines to suggest a new tolerable weekly intake level, as there is no evidence for a threshold for critical lead-induced effects (EFSA, 2010).

Several routes of exposure have been identified for humans. Gaseous lead compounds, for example as produced through combustion of lead petrol, can enter through the respiratory system, while solid compounds can be ingested and accumulate in body through the digestive system.

Chapter 5 – The authors examined genetic distinctness of two endemic subspecies of deer in Korea [Chinese water deer (*Hydropotes inermis argyropus*) in Korean peninsula and Siberian roe deer (*Capreolus pygargus ochracea*) in Korean peninsula and Korean Jeju Island] by nuclear *IRBP* and mitochondrial cytochrome *b* and control region sequence analyses of two species in Asia. Two sympatric phylogroups of Chinese water deer in Korea, previously recognized from their cytochrome *b* and control region sequence analysis, were indistinguishable in this *IRBP* sequence analysis. On the other hand, genetic distinctness of the Siberian roe deer in Jeju Island, previously revealed from their control region sequence analysis, was confirmed from present *IRBP* analysis. The authors concluded that Jeju Island roe deer is an allopatric phylogroup, which differs from other populations in Asia, including Korean peninsula, northeastern China, far-eastern Russia, Mongolia, and western Siberian Kurgan. Each endemic subspecies of deer in Korea is genetically distinct to be considered as evolutionary significant unit (ESU), and each subspecies is unique in their distributional aspect as well. However, conservation status of two deer in Korea is in striking contrast. The former is abundant, and is regarded as a pest, game mammal, whereas the latter was rare and is treated as a protected mammal. In order to determine protection tactics for two subspecies of deer in Korea and other deer from other regions as well, the authors recommend adoption of a criterion, i.e., evaluation of distributional and evolutionary uniqueness in global scale, to upgrade into threatened category, when population size in a certain local area is large enough to fit the IUCN guideline for the least concern category.

Chapter 6 – The Japanese Archipelago is elongated in a north–south direction and contains various vegetation zones. Sika deer (*Cervus nippon*) range in these zones. In northern Japan, deciduous oak forest and beech forest prevail. The floors of these forests are densely covered by dwarf bamboos, which, because they are evergreen, provide important forage for sika deer during winter. The removal of dwarf bamboos by deer sometimes favors seedlings and facilitates forest regeneration. Northern deer undertake vertical migrations according to snow during winter. Winter ranges are therefore highly populated, resulting in the barking of trees. In southwestern Japan, lowlands are covered by evergreen oak forests with dark and poorly vegetated floors. However, the existing understory plants are evergreen, and deer food resources show less seasonal fluctuation than in northern Japan. Southern deer are sedentary and do not show seasonal changes in density. Mountains in southern Japan are covered by evergreen trees, and *Sasamorpha borealis*, a dwarf bamboo, grows on the forest floors. When deer densities are high, as is often the case on small islands, many plants are decimated by deer grazing and are replaced by unpalatable plants or by *Zoysia japonica*, a lawn sward. Recently, sika deer have expanded their range into snowy areas and alpine zones.

In: Deer: Habitat, Behavior and Conservation ISBN 978-1-62100-676-3
Editors: A. A. Cahler and J. P. Marsten

Chapter 1

INFLUENCE OF ANTHROPOGENIC DISTURBANCES ON ACTIVITY, BEHAVIOR AND HEART RATE OF ROE DEER (*CAPREOLUS CAPREOLUS*) AND RED DEER (*CERVUS ELAPHUS*), IN CONTEXT OF THEIR DAILY AND YEARLY PATTERNS

Susanne Reimoser
Research Institute of Wildlife Ecology,
University of Veterinary Medicine, Vienna, Austria

ABSTRACT

Human recreation influences deer in various ways. To better understand these interactions three roe deer and two red deer were investigated during controlled disturbance trials. By means of implanted telemetry transmitters heart rate and positions of the head were recorded. Heart rates and behavior were compared before, while and after confronting the animals with various stimuli. Furthermore heart rate and activity of 14 roe deer and 8 red deer were continually recorded to obtain the normal daily and yearly patterns without disturbance trials. Altogether 6.8 million minutes of data were analysed. For each registered minute a distinction between active and inactive behavior could be made.

Daily activity patterns of roe deer varied greatly within different seasons: From November to June the animals were more active in the daytime, whereas in late summer and early autumn they showed pronounced nighttime activity. In contrast, red deer showed a relative constant ratio of day to night activity. Both roe and red deer were most active during twilight, roe deer more so at dawn, red deer more at dusk. Resting heart rate of red deer was lowest in winter (~44 beats/min), and highest in May and June (~71 beats/min). Roe deer showed less pronounced seasonal differences of resting heart rate (65-76 beats/min). Lunar cycles influenced activity of red deer, but not of roe deer.

Both roe deer and red deer exhibited stronger behavioral reactions ("activation") and heart rate increases towards optical stimuli (walking persons, dogs, horse riders and hang-gliders) than to acoustic stimuli (chain saw, radio music, dog barking) with exception of gunshots, which also elicited high reactions. Activation was associated with daytime and season. Both roe and red deer were more often activated in the morning hours and late afternoon than at noontime. Altogether the reactions of roe deer towards stimuli were short and strong, whereas red deer showed less obvious but persisting reactions. Behavior of red deer was disturbed for at least 10 minutes after gunshots and persons passing by, while horse riders, hang-gliders, tractors and most acoustic stimuli had milder reactions as consequence.

Roe and red deer showed different strategies of compensating after being disturbed. By roe deer the length of resting bouts were shortened on days with disturbance trials, but the number of resting bouts per day increased, so that the total activity budgets for daytime as well as for 24 hours were not influenced by disturbance trials. Red deer were more active during daytime when disturbed without changing their night activity (total daily activity increased), however, they had lower resting heart rates on days with disturbance trials. Recommendations are given to minimize disturbances of deer.

Keywords: human disturbance, hiking, horse riding, dogs, hang-gliders, gunshots, compensation strategies, controlled trials

1. INTRODUCTION

1.1. Objectives

Increasing human recreational activities lead to more and more disturbances of wild animals living in already size-reduced habitats that are additionally dissected by housing developments and traffic infrastructure

(Jaeger et al., 2005; Jiang et al., 2009). The question arises to what extent animals are disturbed by recreationists and other land users such as foresters and hunters and how far their well-being is impaired. Particularly protected land areas should function as sanctuaries for native animal and plant species, but should also enable people to get in touch with nature. To harmonize both aims is one of the difficult tasks for ecologists, biologists and wildlife managers (Burch, 1988; Boyle and Samson, 1985). This study wants to improve the scientific basis for management decisions. The objectives are:

- to measure experimentally the impact of different anthropogenic disturbances on behavior and heart rate of roe deer (*Capreolus capreolus*) and red deer (*Cervus elaphus*);
- to relate the findings to the daily and yearly activity patterns of the deer species;
- to compare the results with the outcomes of other studies to this topic.

1.2. Fundamentals of Researching Stress in Wild Animals

1.2.1. Definition of Disturbance

"Disturbance" is interpreted very differently in literature, either as the cause of an occurrence, as the effect, or as both combined (Rykiel, 1985; Archer and Stokes, 2000). In this study the term "disturbance stimulus" is defined as the cause of changes in behavior, activity and heart rate of the animals and "disturbance effect" as their reaction to the disturbing agent.

Rykiel (1985) characterizes a disturbance effect by the *direction* of the impact, the *magnitude* and the *persistence*. A disturbance effect can either be *transient*, such as a temporary increase in the heart rate or it can be *permanent*, for example a change in the activity pattern of the animal. According to Reichholf and Schemel (1988) a disturbance can be specified by its intensity, its duration and its frequency. Christensen (1988) describes five parameters for disturbance: the *frequency*, the *return time* at a particular site, the *intensity* irrespective of its effect on the animals, its *severity*, i.e. its impact on the animal and its *predictability*. Horejsi (1976) distinguishes between active and passive harassment. Active harassment leads to obvious changes in behavior such as flight, whereas passive harassment shows only subtle changes in activity.

1.2.2. Definition of Stress

Seyle (1936) conceived stress as the nonspecific response of the body to any demand. These demands can be very different in nature: e.g. heat, injuries, hunger or toxins. Moberg (1985) later pointed out that stress reactions are not always nonspecific, but rather various stress factors can affect the same individual differently. Jenkens and Kruger (1973) discern between *psychological* stress (such as fear and frustration) and *somatic* stress (for instance injuries or hunger). Stress reactions depend on the responsiveness of the animal being exposed to the stressor (Porges, 1985). Different individuals can show varied reactions upon the same stressor and even one and the same individual responds dissimilar at different times. The reaction towards a certain stressor is modified by many factors, e.g. age, sex, condition, temperament of the animal, previous experiences or concurrent environmental conditions (Jenkens and Kruger, 1973).

Cannon (1929) was the first to recognize the relevance of the autonomic nervous system in stress procedures. During an "emergency state" triggered by pain or agitation, epinephrine is released and heart rate increases. Further physiological responses are increase of respiration rate, oxygen consumption, blood sugar and body temperature (Gabrielsen and Smith, 1995). Next to the short-term effects of the autonomic nervous system the behavior of the animal plays an important role in stress procedures. Behavioral reactions are often the most simple and economical way for animals to respond towards stress, for example finding a shady spot on a hot day (Moberg 1985).

1.2.3. Relevance of Investigating Stress in Animals

Stress reactions can generally be interpreted as adaption of the organism to changes in the environment. An environment poor in stimuli can at times even have an adverse effect (Fraser et al., 1975). But usually an excess of excitation is detrimental for the animal, yes, can even endanger its life (Geist, 1971). This is especially the case when more than one stress factor affect the animal simultaneously, e.g. harsh weather and harassment.

Due to the increased release of epinephrine the basic metabolic rate is raised, which leads to an increased energy consumption. Furthermore, the time the animal spends with forage intake is reduced when disturbed frequently, as could be observed e.g. in chamois (Cederna and Lovari, 1985) and red deer (Jayakody et al., 2008). Manor and Saltz (2003) observed that gazelles distinguished between nuisance disturbances, such as by recreationists, and direct threats to survival, such as feral dogs. Yet also nuisance disturbances may be detrimental to the animals, resulting in increased energy expenditure and vigilance at the expense of foraging activity.

Animals may respond to disturbance from humans by avoiding areas of high risk (Gill et al., 1996; Hewison et al., 2001). Yet animals with no suitable habitat nearby may be forced to remain despite the disturbance (Mysterud et al., 1999), even if survival or reproductive success is reduced (Gill et al., 2001). Sometimes animals temporally shift into areas with unfavorable food supply, e.g. from open grassland into woodland as with chamois (Rauer-Gross et al., 1988), wapiti (Hernández and Laundré, 2005), or red deer (Jayakody et al., 2011). Particularly in winter this can lead to a negative energy balance, in which the animal expends more energy per time unit than it can ingest. Thus survival in winter is jeopardized (Geist, 1971). Also the digestion of ruminants is disordered when the animals are frequently disturbed, since sufficient resting periods for ruminating are necessary. With reduced chewing less saliva is produced inducing acidosis of the rumen. This leads to poorer condition of the animals (Barth, 1984).

Further negative effects of disturbance are reduced fertility and abortion. Squibb et al. (1986) observed a sudden decline in conception frequency in wapiti at the beginning of the hunting season. They assume an interference of the rut or early embryonic mortality. It is documented for reindeer that chasing heavily pregnant females through deeply crusted snow can precipitate abortion (Geist, 1971). Moving away into unfavorable calving areas results in higher mortality of young (Whitten et al., 1992).

In some cases wild animals migrate in other regions due to anthropogenic disturbances. For example wapiti were more easily driven off from their home range by firewood gatherers than the less sensitive mule deer (Ward, 1984) and wapiti also avoided a newly built road (Czech, 1991). White-tailed deer could be dispelled from their home range through high hunting activity (Sparrowe and Springer, 1970) and traditional winter habitats of red deer remained unused after high hunting pressure (Schmidt, 1992). Female fallow deer were more apt to abandon habitats frequently disturbed, whereas males tended to remain in these areas (Apollonio et al., 2005).

Yet usually cervids are philopatric and return to their home range after the disturbance has ended. According to Sweeney et al. (1971) white-tailed deer chased by dogs out of their home ranges usually returned within one day. Whereas male white-tailed deer migrated away when often disturbed through cattle drives, females showed higher fidelity to their home range (Hood and Inglis, 1974). Even after repeated targeted disturbances red deer would not abandon their home range (Georgii, 1980). Jeppesen (1984) observed that roe deer stayed within their home range under cover or left it only for a short time when disturbed by orienteering and driven hunts. After an extensive forest fire red deer returned to their home range despite the totally changed landscape and lacking forage (Raesfeld and Reulecke, 1988). Also caribous which were often chased and harassed by helicopters would not leave their home ranges (Bergerud et al., 1984).

Not only wild animals suffer through excessive disturbances, also the forest is indirectly damaged through bark peeling and browsing of young shoots. Two main reasons are relevant. On the one hand the animals seek hiding cover in the forest, but cannot find appropriate forage. Secondly, peeling can prevent rumen acidosis, in that more alkaline saliva is produced through the chewing motion (Onderscheka, 1979). Especially in winter harassment of deer can provoke bark peeling (Reimoser et al., 1987). How and to which extent disturbances of wild ungulates lead to game damage is poorly understood and should be more investigated (Völk and Gossow, 1997; Reimoser, 2005).

Overviews of effects of various disturbance stimuli on wildlife can be found in Knight and Gutzwiller (1995), Ingold (2005), Stankowich (2008) and Snetsinger (2009).

1.2.4. Measuring Stress

Many different physiological parameters can serve as indicator for stress and agitation (Levine, 1985; Pfaff, 1982). Some of them though are unsuitable for the investigation of targeted disturbance trials on wild animals. Measuring blood pressure, skin temperature, electrodermal activity, muscle tone, EEG and oxygen consumption is technically difficult. The concentrations of catecholamines, free fatty acids and corticosteroids in blood change with impact of stressors, yet since the capture and the taking of blood samples causes strain in wild animals, these parameters must also be discarded. Besides this, interpreting the concentration of corticosteroids in blood plasma is extremely difficult, since they can elevated or lowered depending on whether the situation is acute or chronic (Stott, 1981; Rushen, 1991). The isozyme LDH-5 proved to be a suitable parameter for measuring stress in fallow deer (Jones and Price, 1990), but is also associated with the problem of taking blood samples as above-mentioned. Cortisol levels can also be measured in faeces (Huber et al., 2003; Keay et al. 2006). This evades the problem of capturing the animal, but it is difficult to allocate the samples to an individual wild animal and if rendered possible it is still impractical to relate the variations in cortisol level to specific short-term disturbances (von der Ohe and Servheen, 2002).

Another characteristic connected to stress is the enlargement of the adrenal glands. Werner (1976) could find considerably enlarged adrenal glands as only pathological change in a roe buck which died during disturbance trials. Yet the measurement of the adrenal glands cannot give information about differing effects of various stress factors on the same animal and about the impact of short-term disturbances. Furthermore the individual variability of adrenal weights in roe deer is considerable (Georgii, 1976).

Heart rate in contrast is a very suitable parameter for this study. By collecting the data via telemetry the behavior of the animals is not additionally affected. Above all short-term reactions can also be measured. Harlow et al. (1987) state that remote monitoring cardiac frequency can be used to predict adrenal function as they found a linear relationship between heart rate and blood cortisol in bighorn sheep. The correlation of heart rate and energy consumption is also relevant, as was verified in the redshank (Ferns et al., 1979), caribou (Fancy and White, 1985) and white-tailed deer (Holter et al., 1976; Mautz and Fair, 1980). According to Freddy (1979) heart rate gives information on psychic strain of the animal even if the animal doesn't show noticeable behaviorial reactions. Ferns et al (1979) observed a doubling of the heart rate in sleeping redshanks as soon as a person entered the laboratory, while the birds didn't change their body position. Ball and Amlaner (1979) had similar results with herring gulls. Jacobsen (1979a) measured an increase of 50% of the normal heart rate in lying white-tailed deer calves, when alarmed. The heart rate was 2.4 times higher in moose and 2.7 times higher in caribou when agitated (Roshchevskii et al., 1976), and Freddy (1979) could also observe an increase in the heart rate of mule deer when persons passed by.

Arousal of animals does not necessarily lead to an increase of heart rate. Under certain circumstances the heart rate can also decrease. Red deer calves (Espmark and Langvatn, 1979), white-tailed deer calves (Moen et al., 1977; Jacobsen 1979b) as well as eastern cottontail rabbits (Smith and Worth, 1979) showed the so-called "alarm bradycardia". Normally this is the case with animal species that show the behavior of lying prone when alarmed ("freezing and hiding" in opposition to "fight and flight"). Stöhr (1982) observed that individual tupajas lowered their heart rate while other individuals increased it when confronted with the same stimulus. Animals that normally reacted with tachycardia could also show bradycardia when confronted with a very intense stimulus. Miller (1982) interprets these two contrary reactions as two different inherent behavioral patterns when in fear. But also animals without freezing behavior when alarmed can react with bradycardia when they are alert, e.g. wolves (Kreeger, 1990) and bighorn sheep (Stemp, 1983). Experiments analysing the reaction time of humans confirm the assumption that the heart rate slows down when the test person is attentive and in a state of anticipation (Zimmer et al., 1989).

Besides the so far mentioned examples of short-term ("phasic") changes in heart rate through disturbances, also long-term ("tonic") changes in heart rate can occur, as Stöhr (1982) could provide evidence in tupajas.

In addition to analysing heart rate, the monitoring of behavior is suitable for examining stress in wild animals. Both short reactions and changes in activity patterns in the long run offer valuable clues to the agitation of the animals. Particularly, because the heart rate is not only a criterion for the psychic status of the animal, but also strongly influenced by its activity and therefore insufficient as a sole measure (Stemp, 1983), it is important to include the behavioral composition and the overall activity before and after disturbance trials.

Berger et al. (2002) suggest analysing how far ultradian rhythms in activity patterns are synchronized with the external daily cycle as measure for stress. This was shown to be an indicator for drastic impacts on the animal, such as chases or immobilization, but it remains to be seen if disturbances on a fine-scale can also be detected.

Another approach to investigate habitat use and general activity patterns depending on disturbance impacts is the use of collars with GPS and activity sensor (e.g. Sibbald et al., 2001; Adrados et al., 2003; Sunde et al., 2009; Naylor et al. 2009; Duscher et al., 2009). Contrary to direct observation of flight distances, animals responding beyond the perceptive range of the observer can also be detected via GPS (Preisler et al., 2006).

To draw conclusions about the wellness of an animal, numerous parameters, such as health, fertility, physiological measurements and behavior should be dealt with (Blackshaw, 1986). Heart rate is only one aspect, however combined with behavioral monitoring a good indicator for disturbance of wild animals is available.

1.2.5. Habituation and Sensitization

If an identical stimulus is applied repeatedly the reaction of an individual toward this stimulus lessens, the animal habituates (Thompson and Spencer, 1966). Yet an animal's behavior is not simply habituated or non-habituated, but a matter of degree, and responses occur in differing magnitudes in different contexts (Whittaker and Knight, 1998).

An extraneous stimulus can disrupt the process of habituation, this is called dishabituation. If the stimulus repeatedly applied has an aversive character, the opposite of habituation can occur, the animal is sensitized towards this stimulus (Groves and Thompson, 1970).

Espmark and Langvatn (1985) could observe habituation as well as dishabituation regarding alarm bradycardia when confronting red deer calves with persons passing, stones thrown and dog barking. Gaisbauer (1988) noticed that the flight reaction of roe deer decreased when approached repeatedly. Albrecht (1988) differentiates between unspecific and person-specific habituation. After three months he could approach a group of chamois approximately 27 meters, whereas the flight distance towards other persons was in average 46 m. Altmann (1958) observed that moose had a shorter flight distance towards frequently encountered fishermen than to other persons.

But also the spatial arrangement plays an important role in habituation towards stimuli. When Schleidt and Shalter (1983) let dummies pass angel fish repeatedly from the same direction the first strong flight reactions declined. If the dummy was pulled from a different direction the original strong reaction took place again. Shalter (1984) came to similar results with domestic and jungle fowl. He deduced that animals in a natural environment don't habituate towards predators contrary to the simulated experiments, since these always approach their prey from different, unaccustomed directions.

Miller and Gunn (1984) report that individual muskoxen habituate towards helicopter overflights while others are sensitized, and conclude that the tolerance threshold towards stimuli can be individually different. Moen et al. (1982) could not find any habituation of white-tailed deer towards snowmobiles, and MacArthur et al. (1982) noticed higher heart rate reactions when approached by persons repeatedly. This is in accordance with the observations of Freddy et al (1986), as they measured progressive higher heart rates in mule deer from the first to third approach. Contrary to his expectations Stemp (1983) also noticed a sensitization of heart rate reactions in bighorn sheep when repeatedly approached by humans. From this he concluded that frequent disturbances lead to sustained heart rate reactions.

In summary it can be said that animals react differently towards regularly recurring disturbances. Mostly they gradually habituate, but they can also show increased sensitivity to disturbing factors.

1.2.6. Vigilance

Potential prey can protect themselves from predators in being vigilant. The time spent in vigilance goes at the expense of other behavioral elements, which are vital for the animal such as forage intake (Turner, 1979; Frid and Dill, 2002). An alternative to reduce the vigilance time for the individual animal without reducing its security is to herd together. Turner (1983, cit. Kurt, 1991) compared the time budget of single roe deer compared to roe deer in groups. Detached animals spend about 26 minutes per hour in vigilance behavior, deer living in larger groups only about 12 minutes. Büttner (1980) and Lark and Slade (2008) could also observe a decrease of vigilance behavior for each individual with increasing group size, combined with time gained for forage intake.

Another alternative to lessen the time dedicated to vigilance behavior is in the use of terrain rich in hiding cover (Dimond and Lazarus, 1974). Whereas species in open fields often herd, woodland species respond to disturbances by hiding (Putman, 1988). According to Petrak (1996) red deer show three different strategies when disturbed: seeking hiding cover (mostly at daytime), increasing distance to the source of disturbance (mostly at night), or gaining overview. If hiding cover is sparse, e.g. in winter, it is also feasible for the animal to shift its activity into the night. This matches the observations of Guthörl (1987) and Jeppesen (1989) for roe deer, Beier and McCullough (1990) and Kammermeyer and Marchinton (1977) for white-tailed deer and Jeppesen (1987) for red deer. Red deer studied by Georgii and Schröder (1981) showed evenly spaced activity bouts at daytime in terrain rich in hiding cover, but were mainly nocturnal if hiding cover was sparse. They conclude that the activity pattern of an animal is also regulated by habitat structure.

1.3. Activity Patterns

Activity patterns of animals are influenced by many factors. On the one hand endogenous factors, such as the momentary metabolic state and hereditary behavioral programmes, play an important role, on the other hand exogenous factors, such as weather, day length or disturbances. Silver et al. (1969) did research on the basic metabolic rate of hungry white-tailed deer in respiratory chambers. Despite constant temperature in the chamber the metabolic rate was much higher in summer (May to August) than in winter. This change was ascribed to changing endocrine activity triggered by day length.

1.3.1. Seasonal Activity Patterns

Activity is strongly influenced by season and daytime. In most investigations cervids were far less active in winter than in summer (Georgii and Schröder, 1978; Lieb, 1981; Green and Bear, 1990). Moreover the active and inactive bouts are longer in winter than in summer, where more frequent changes in activity occur (Cederlund, 1981; Jeppesen, 1989). According to Cederlund (1981) this is due to supply of forage. In summer with plenty of forage the rumen is quickly filled and the animal soon reclines to ruminate. The high-energetic, easily digestible food is quickly metabolized, the rumen is soon empty and the animal rises again to forage. In winter the expenditure of time to intake food is much longer, so the activity bouts are prolonged ("rumen fill theorie"). The forage is rich in crude fibers, requiring more time to chew the cud and leading to longer resting bouts.

In opposition to the above-mentioned findings, it was experimentally shown with roe deer (Ellenberg, 1978) and white-tailed deer (Ozoga and Verme, 1970) kept under steady conditions with constant food composition, that they also minimized food intake and activity in winter. This is an indication of a genetic fixation of the seasonal pattern.

Female red deer investigated by Georgii (1981) showed, contrary to roe deer, shorter, but more frequent activity bouts in winter. Wapitis were more nocturnal in winter according to Green and Bear (1990), whereas the same animal species investigated by Lieb (1981) were more active at daytime during the winter.

1.3.2. Circadian Rhythms

Bimodel daily activity patterns, e.g. intensified activity at dawn and dusk and an extensive inactive phase in the early afternoon, are very widely spread in many animal species (Aschoff, 1966). This rhythm is triggered by an endogenous, self-sustaining oscillation, as Aschoff (1966) could show in greenfinches, yet afterwards modified by so-called synchronizers or "zeitgebers". The environment can inhibit or emphasize parts of the activity pattern, but cannot generate a basic rhythm (Ozoga and Verme, 1970). Many investigated deer species showed a bimodal daily pattern (for red deer: Bützler, 1974; Georgii and Schröder, 1978; Georgii, 1981; for wapitis: Collins et al., 1978; Green and Bear 1990; Lieb, 1981; Lieb and Les Marcum, 1979; for white-tailed deer: Kammermeyer and Marchinton, 1977; for roe deer: Cederlund, 1981; Jeppesen, 1989; Mertens, 1984). According to Jeppesen (1989) the bimodal pattern of roe deer is more pronounced in winter, the same is valid for wapitis (Green and Bear, 1990) and red deer (Georgii, 1984).

Turner (1980) ascribes the crepuscular behavior to a better synchronization of the individual animals, since the change in light functions as "zeitgeber". With the individual variable digestion time the behavior desynchronizes at daytime.

2. Material and Methods

2.1. The Deer Investigated

14 roe deer and 8 red deer were investigated for their activity patterns and heart rate (Table 1).

Table 1. Details of study animals for activity patterns

No.	species	sex	age (years) when implanted	enclosure type[1]	trans-mitter type[2]	collected data (days)	heart rate data (days)	high quality activity data (days)[3]
1	roe deer	f	2	1,2	t	332	302	292
2	roe deer	f	2	1	h	35	30	29
3	roe deer	f	3	1,2,3	t	337	210	214
5	roe deer	f	2	1	h	57	13	13
7	roe deer	f	2	1	h	381	338	325
21	roe deer	m	3	1	h	133	116	110
22	roe deer	m	3	1,2,3	t	314	176	173
24	roe deer	m	?	1	h	67	35	29
27	roe deer	m	3	1	t	119	110	102
28	roe deer	m	4	1	h	18	15	14
29	roe deer	m	3	3	t	125	75	73
31	roe deer	m	2	1	h	261	179	185
32	roe deer	m	3	1	h	260	230	229
33	roe deer	m	1	1	h	141	135	127
102	red deer	m	1	4	r	914	325	190
106	red deer	f	3	4	r	1185	141	469
107	red deer	f	3	4	r	1155	185	337
113	red deer	f	2	4	r	1066	703	741
121	red deer	m	10	4	r	791	216	491
130	red deer	f	2	4	r	394	39	171
135	red deer	m	5	4	r	702	33	81
140	red deer	f	1	4	r	913	0	359

[1] enclosure type: 1=50m^2, 2=3500 m^2, 3=46ha, 4=37ha

[2] transmitter type: h = implanted transmitter for heart rate, t = transmitter for heart rate including tilt switch, r = repeater system

[3] high quality data for activity are days with less than 60 minutes missing values

Of 9700 days of collected data, 4754 days (2839 days for red deer, 1915 days for roe deer) were considered as high quality data for activity analysis, with less than 60 minutes missing values per day and good signal strength. 3606 days were used in heart rate analysis.

Three roe deer and two red deer were investigated during the planned disturbance trials (Table 2). The animals were born in enclosures, but were not tame. To avoid interactions between the two species the trials were carried out at separate periods.

Table 2. Details of study animals for disturbance trials

No.	species	sex	age (years)	weight (kg)	disturbance trials (season)
1	roe deer	female	2	20	Sep.10 – Nov.18
22	roe deer	male	3	25	Sep.16 – Nov.18
3	roe deer	female	3	20	Oct.26 – Nov.18
61	red deer	male	1½	75	Dec.10 – Mar.30
51	red deer	fcmale	½	45	Dec.11 – Mar.30

The animals for the disturbance trials were fitted with telemetry transmitters that were able to measure heart rate and the position of the head (up/down). Technical details of the devices are available in Schober et al. (1982). The transmitters were implanted on the left body side between peritoneum and the muscles of the abdominal wall and connected with two subcutaneous ECG-electrodes in the thoracal area and a subcutaneous tilt switch in the neck area (Figure 1, left). The cables to the electrodes and the tilt switch as well as the antenna ran subcutaneously.

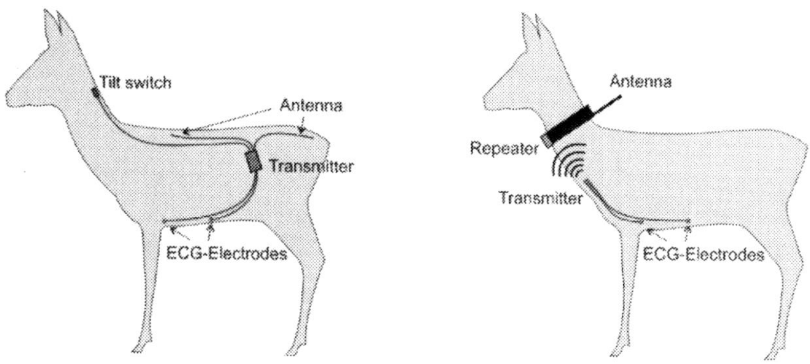

Figure 1. Left: Implanted Transmitter with tilt switch and ECG-electrodes; Right: Repeater system with implanted transmitter, ECG-electrodes and repeater neck collar.

Heart rate was transmitted by sending an impulse of several milliseconds for each QRS-complex of the ECG. The length of the impulse was modulated by the tilt switch depending on whether the head was up or down. The received signals were continuously recorded by two computers. The received field strength was dependent on the position of the implanted transmitters to the receiver and gave information on movements of the animals.

For the general activity and heart rate patterns only 5 roe deer had tilt switches, which allowed identifying feeding while active and "sleep" phases while inactive. The red deer had a repeater system with neck collars carrying larger batteries necessary for transmitting the signals for farther distances (Figure 1, right). With this system the transmitter implanted could be smaller since it only needed the energy to send signals to the neck collar, from where the signals were sent to the receiving station. Details about the repeater system can be found in Schober and Fluch (1995) and Giacometti et al. (2001).

2.2. Enclosure for Disturbance Trials

The disturbance trials were carried out in a 3500m^2 large enclosure (Figure 1) located in the research preserve of the Research Institute for Wildlife Ecology, Vienna. It encloses a meadow with a slight north-facing slope next to a mixed forest (beech-oak-pine) and is situated near the institute buildings.

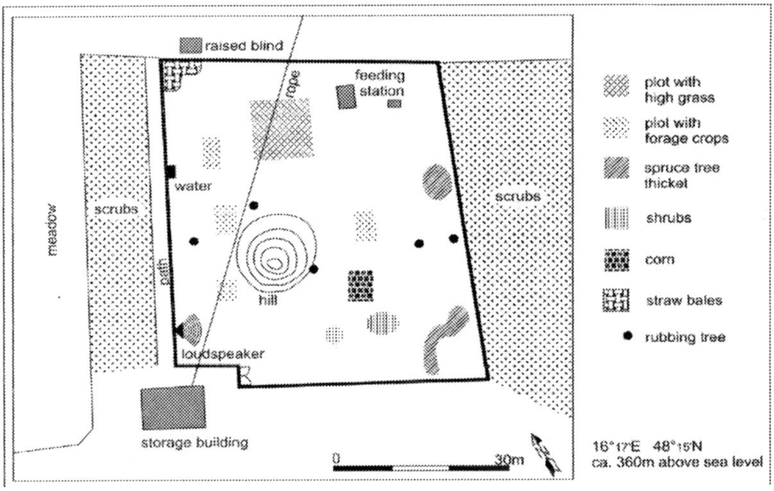

Figure 2. Enclosure.

The meadow was mown twice a year except for one larger patch of high grass and some small patches planted with forage crops (alfalfa, corn, rape, collard greens). In addition small groups of shrubs and young spruce trees provided hiding cover. Videos could be made from the window of a storage building near to the enclosure. A rope for the hang-glider was spanned across the enclosure 4-5m above ground level.

2.3. Disturbance Trials

Different standardized disturbance stimuli were applied in random order with at least half an hour time intervals in between. Heed was paid that each type of stimuli was placed at different times of the day. Subsequent the stimuli and the number of experiments carried out for roe and red deer respectively:

Optical stimuli

- round tour of a single person: One person walks along the fence inside of the enclosure (duration 3-4 minutes; n = 29/32)
- round tour of a person with dog: One person walks with a dog on leash along the fence inside of the enclosure (duration 3-4 minutes; n = 17/17)
- round tour of a group of people (Figure3): at least three persons walk along the fence inside of the enclosure talking at normal sound intensity (duration 4-5 minutes; n = 12/14)
- round tour with a horse (Figure 3): one person rides either a white or brown horse in slow tempo or in trot and gallop along the fence inside of the enclosure (duration 2-4 minutes; n = 19/31)
- round tour with a group of horses (Figure 3): three horses are ridden in slow tempo along the fence inside of the enclosure, with riders talking to each other (duration 4-5 minutes; n = 5/10)
- hang-glider (Figure 3): a model of a hang-glider (1.5m wing span) glides about 4-5 m above the ground on a rope (duration 2-3 minutes; n = 10/14)
- tractor: a small tractor is driven outside the enclosure next to the fence (duration 2-3 minutes; n = 10/14)
- ascent of raised blind: one person climbs a ladder up to a raised blind next to the enclosure (duration 1-2 minutes; n = 12/-)

- feeding: the animals are fed by the same person daily at 3 p.m. (duration 3-5 minutes; n = 31/29)

Optical and olfactory stimuli

- fire: a straw bale is set to fire inside of the enclosure (duration 8-10 minutes; n = -/5)

Acoustical stimuli

- radio music: music is played for 3 minutes from a loudspeaker fixed to the fence of the enclosure (n = 12/12)
- dog barking: a sequence of 20 seconds barking, 30 seconds pause and 40 seconds barking is played from a tape recorder (n = 6/12).
- chain saw: a chain saw is turned on about 20 m from the enclosure for 3 minutes (n = 11/11)
- gun shot: a cracker is lit near the enclosure (n = 24/17)

With all acoustical disturbance trials care was taken that the animals could not perceive the person conducting the trial.

Figure 3. Disturbance trials. Above left: group of persons, above right: hang-glider, below left: single horse rider, below right: group of horses riders.

2.4. Data Acquisition and Interpretation of the Data

Video recordings made during the disturbance trials, as well as the stored telemetry signals showing field strength, head position and heart rate of the animals were drawn upon for the investigation. The received telemetry data were aggregated to entities of one minute and stored on magnetic tapes ("offline data") and were the basis for long-term analysis. Simultaneously non-aggregated data from heart beat to heart beat could be observed on the computer screen and were also stored ("online data"). These data were very suitable for short-term investigations when analysed in playback. With the high temporal resolution each single heart beat could be reconstructed.

2.4.1. Analysis of the Long-Term Recordings

The offline-data gave information about the activity of the animals over a long period of time. For every minute it could be discriminated if the animal was active or inactive. Inactive was defined as the animal lying, all other behaviour types were classified as active. For each individual an actogram was plotted showing active and inactive phases in a resolution of one minute.

In the analysis of the offline-data, the animal was considered active when the field strength changed often indicating movements of the animal. In addition, the heart rate was higher than in inactive animals, and often the head position changed (Figure 4 and 5).

Day: 26.1.
Hour: 22

Minute	9	10	11	12	13	14	15	16	17	18	19	20	21	22
heart frequency: mean	82	82	86	96	98	98	92	94	96	94	92	92	94	92
heart frequency: maximum	90	92	110	112	106	110	108	108	110	108	104	106	110	114
heart frequency: minimum	74	74	76	86	88	84	80	78	88	80	86	82	80	76
head changes (up/down)	---	2	4	4	---	2	---	---	---	---	---	2	---	4
head down (sec)	60	49	42	41	60	59	60	60	60	60	60	59	60	48
max. contin. head down time	60	44	18	28	60	47	60	60	60	60	60	48	60	26
max. contin. head up time	---	11	10	13	---	1	---	---	---	---	---	1	---	8
number of signal adjustments	---	---	9	12	21	6	---	---	4	5	6	18	11	17
signal strength: mean	54	54	55	55	46	53	53	54	56	54	54	52	48	55
signal strength: maximum	54	54	58	62	60	57	53	54	57	56	55	58	57	59
signal strength: minimum	54	54	44	47	31	48	53	54	52	48	47	47	39	46

Figure 4. Example of an offline-file of an animal in active state. The long phases with head down and signal adjustments in middle range are typical for browsing or olfactory control of the ground surface.

Day: 27.1.
Hour: 21

Minute	43	44	45	46	47	48	49	50	51	52	53	54	55	56
heart frequency: mean	98	142	120	176	138	160	172	118	98	116	162	166	176	124
heart frequency: maximum	130	308	150	172	172	288	368	136	112	146	410	308	304	206
heart frequency: minimum	82	86	100	116	116	124	120	102	90	90	104	146	130	102
head changes (up/down)	9	8	2	6	11	9	12	4	4	10	9	11	16	5
head down (sec)	12	8	8	6	17	10	19	25	9	10	3	7	11	5
max. contin. head down time	4	2	8	2	5	4	8	23	8	4	1	3	3	3
max. contin. head up time	19	26	46	26	13	19	20	19	21	22	32	22	30	24
number of signal adjustments	26	32	10	36	31	35	21	11	6	37	28	39	7	9
signal strength: mean	57	32	10	36	31	35	21	11	6	37	28	39	7	9
signal strength: maximum	98	68	40	96	90	99	54	53	47	97	62	97	52	48
signal strength: minimum	42	44	29	48	28	40	39	39	39	38	38	41	38	31

Figure 5. Example of an offline-file of an animal during locomotory activity. Heart rates are high and signal strength is frequently adjusted.

Day: 28.1.
Hour: 19

							↓							
Minute	28	29	30	31	32	33	34	35	36	37	38	39	40	41
heart frequency: mean	74	72	78	82	80	76	88	100	104	106	104	88	88	94
heart frequency: maximum	84	82	96	108	96	82	136	136	124	124	114	114	102	112
heart frequency: minimum	70	68	68	70	70	68	74	68	74	92	94	76	74	84
head changes (up/down)	---	---	---	---	---	---	1	3	6	3	1	4	9	3
head down (sec)	---	---	---	---	---	---	9	10	30	32	38	20	21	11
max. contin. head down time	---	---	---	---	---	---	9	9	16	19	38	10	14	7
max. contin. head up time	60	60	60	60	60	60	51	30	18	21	22	25	12	42
number of signal adjustments	---	---	---	---	---	---	9	10	21	1	5	2	3	16
signal strength: mean	51	51	53	50	53	50	53	55	54	46	46	33	32	31
signal strength: maximum	51	51	53	50	53	50	62	57	59	100	51	34	35	41
signal strength: minimum	51	51	53	50	53	50	53	48	46	46	37	32	30	17

Figure 6. Offline-data of an animal standing up at 19:34.

Typically when browsing the head is down most of the time (Figure 4), while very high heart rates and many changes in field strength are characteristic when the animal is running (Figure 5). The transition towards an active phase was indicated by changes in field strength, changing head positions and a marked increase in the mean heart rate. After a while the heart rate dropped to a lower level, but still stayed higher than the rate in the inactive phase prior to standing up (Figure 6). When changing from active to inactive (lying down) the heart rate decreased gradually (Figure 7). In the inactive state changes in field strength are minimal and the heart rate is steady and low. Few head changes can occur when the animal is grooming itself or browsing while laying, yet the heart rate stays very low (Figure 8). “Sleeping” is defined as an animal lying with its head on the ground. In this state the heart rate decreases even more (Figure 9).

Day: 26.1.
Hour: 17

Minute	42	43	44	45 ↓	46	47	48	49	50	51	52	53	54	55
heart frequency: mean	94	98	88	88	76	74	74	72	74	72	72	72	72	72
heart frequency: maximum	192	112	104	110	82	84	78	76	86	76	76	78	76	76
heart frequency: minimum	80	84	82	76	72	70	68	68	66	68	70	66	70	70
head changes (up/down)	8	7	4	2	---	---	---	---	---	---	---	---	---	---
head down (sec)	26	39	23	23	---	---	---	---	---	---	---	---	---	---
max. contin. head down time	10	25	20	23	---	---	---	---	---	---	---	---	---	---
max. contin. head up time	21	11	26	26	60	60	60	60	60	60	60	60	60	60
number of signal adjustments	25	8	6	16	---	2	---	---	---	---	---	---	---	---
signal strength: mean	58	58	58	51	42	36	32	35	35	35	35	35	35	35
signal strength: maximum	62	60	59	58	42	100	32	35	35	35	35	35	35	35
signal strength: minimum	51	51	50	43	42	33	32	35	35	35	35	35	35	35

Figure 7. Offline-data of an animal lying down at 17:45.

Day: 7.9.
Hour: 15

Minute	40	41	42	43	44	45	46	47 ↓	48	49	50	51	52	53
heart frequency: mean	66	64	66	68	64	64	66	68	62	60	60	62	74	60
heart frequency: maximum	70	70	72	76	72	68	72	120	66	68	70	78	96	68
heart frequency: minimum	62	56	64	62	60	62	58	58	74	54	52	58	60	56
head changes (up/down)	---	---	---	---	---	---	---	1	1	2	4	3	4	2
head down (sec)	---	---	---	---	---	---	---	2	50	55	46	34	8	44
max. contin. head down time	---	---	---	---	---	---	---	---	50	37	28	26	6	44
max. contin. head up time	60	60	60	60	60	60	60	58	10	5	10	22	25	8
number of signal adjustments	---	---	---	---	---	---	1	5	---	---	---	---	---	---
signal strength: mean	52	51	54	51	52	52	57	53	53	54	54	54	54	54
signal strength: maximum	52	51	54	51	52	52	54	53	53	54	54	54	54	54
signal strength: minimum	52	51	54	51	52	52	---	45	53	54	54	54	54	54

Figure 8. Offline-data of a resting animal. At 15:47 there are slight adjustments in signal strength and some head changes. The heart rate remains at a low level. This can be interpreted as browsing or grooming in resting posture.

Day: 5.9.
Hour: 3

Minute	41	42	43	44	45	46	47	48	49	50	51	52	53	54
heart frequency: mean	92	86	82	78	82	82	80	80	80	84	78	76	76	74
heart frequency: maximum	96	94	88	84	86	92	90	92	86	128	86	80	86	98
heart frequency: minimum	84	76	78	68	76	74	70	74	76	72	68	72	70	66
head changes (up/down)	---	---	---	---	---	---	---	---	---	1	---	---	---	---
head down (sec)	---	---	---	---	---	---	---	---	---	46	60	60	60	60
max. contin. head down time	---	---	---	---	---	---	---	---	---	46	60	60	60	60
max. contin. head up time	60	60	60	60	60	60	60	60	60	14	---	---	---	---
number of signal adjustments	---	---	---	---	---	---	---	---	---	---	---	---	---	---
signal strength: mean	55	55	55	55	55	55	57	54	55	55	55	55	55	55
signal strength: maximum	55	55	55	55	55	55	57	54	55	55	55	55	55	55
signal strength: minimum	55	55	55	55	55	55	57	54	55	55	55	55	55	55

Figure 9. Offline-data of a resting animal laying its head on the ground ("sleep") at 3:50.

The interpreted offline-data was compared with synchronous video recordings of three roe deer and 4 red deer. The percentage of falsely classified minutes was between 1.4% and 4.0% for the female roe deer. These errors occurred almost always with short active or inactive bouts under five minutes. Falsely active were situations when the animal lay down and stood up again shortly afterwards. Falsely inactive were situations when the animal stood motionless several minutes (for example standing alert), since the field strength remained unchanged, the head was incessant up and the heart rate was unusually low (alarm bradycardia). The roe buck often stood still for long times, instead of lying down, but showed all signs of inactive behavior, sometimes ruminating while standing. In 13.2% of the offline-data classified as inactive the buck was standing, thereof 40.2% standing and ruminating and 40.2% standing alert.

Estimating the number of activity changes (from active to inactive and vice versa) misclassifications ranged between 11.1% and 16.7%. These errors resulted from short active or inactive phases under five minutes length that weren't identified. In relation to the total daily activity time these errors could be neglected, yet other parameters such as number of activity bouts or mean duration of activity bouts were influenced.

With red deer the falsely classified minutes were 3.7% and 4.7% for two animals observed under controlled conditions in a smaller enclosure (n=14326 minutes). Food intake was falsely classified in 5.5% of all minutes for animal No. 130 and 18.4% for No. 106. The high values for No. 106 is based on the fact that the head was more up than down while eating from a feeding station in the smaller observation enclosure. It can be expected that the error ratio is less high when the animal is eating grass at ground level which was the case in the large enclosure. From random video recordings (n=480 minutes) in the large enclosure falsely classified minutes lay between 1.1% and 6.7% for three animals. A source of error was also with the dominant male stag (No. 121). In one video observation the old stag lifted his head only twice in 1½ hours during browsing by reason that he was accompanied by a younger stag that took over the vigilant behaviour necessary. In such cases the old stag was sometimes falsely classified as being inactive.

To compare daytime and nighttime activity the time span from one hour after sunrise until one hour before sunset was defined as daytime, and one hour after sunset until one hour before sunrise as nighttime. On days with more than 5% missing values (no signal received or signal could not be interpreted correctly) the day was not evaluated for activity. For analysis of bout length only phases with no preceding or following missing values were incorporated. Also the first nine days after capturing and handling the animals were omitted. Such procedure is also recommended by Morellet et al. (2009).

2.4.2. Analysis of the Short-Term Recordings

The online data were suited for making statements about short-term changes of heart rate after applying disturbance stimuli. Both the magnitude of heart rate changes and the time span until the heart rate drops back to the resting values are indicators of the intensity of agitation for the animals. The moment the heart rate attained more than twice the standard deviation of the mean heart rate before the disturbance trial was defined as the outset of the disturbance effect. The ending of the disturbance effect was defined as the moment the heart rate dropped and remained at least one minute between the thresholds of double standard deviation before the trial (Figure 10).

The disturbance intensity was defined as the number of additionally performed heart beats compared to undisturbed behavior and was computed from the area compromising the heart rate curve from outset to ending of the disturbance effect and the mean heart rate before the disturbance trial began (Figure 10). The intensity of disturbance is particularly important since it is closely connected to the energy expenditure of the animals. Cases with animals changing from inactive to active behavior after the disturbance trial were omitted from analysis since it was not possible to determine the exact ending of the disturbance effect, because heart rate shifted to a higher level when active. The heart rate would fall back under the original threshold only after lying down again, which could be hours later (Figure 10, right diagram).

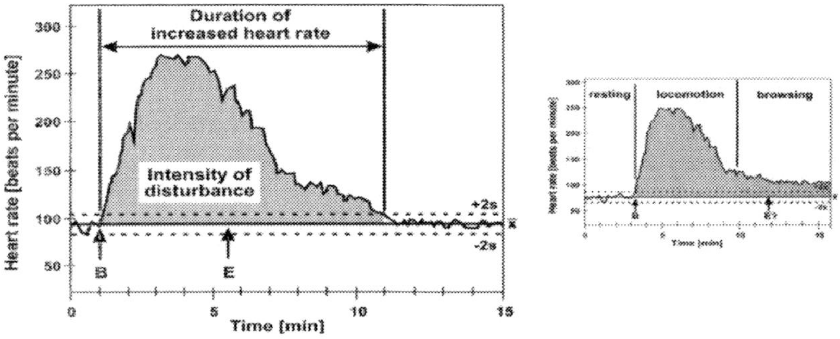

Figure 10. Model of disturbance intensity with heart rate as criterion: heart rate before, during and after a disturbance trial; B = beginning of trial; E = end of trial; shaded area = number of additionally performed heart beats. Large model on left hand shows a situation with the animal in the same state (active or inactive) before and after the disturbance. Small model on the right shows a situation with an inactive animal changing to active behavior after the disturbance.

2.4.3. Analysis of Video Recordings

The animals were video-recorded during the disturbance trials as well as five minutes (roe deer) to ten minutes (red deer) before and after the trials. An inventory of all exhibited actions of the animals (ethogram) was compiled (Figure 11) and used as the basis for the analysis of the video recordings. The beginning of each change in behavior was written down and the duration to the next change in seconds calculated. Since the clockwork of the videorecorder was daily matched to the computer, the actual behavior could be correlated with the corresponding heart rate.

48 different behavior types were identified, which were later on aggregated to 7 main categories:

- *Lying:* All behavior types in which the animal is lying on the ground are included: lying with and without ruminating, lying with the head on the ground, lying rolled up, while grooming, browsing or checking the surroundings with the sense of smell.
- *Standing:* The animal stands without vigilance behavior. Standing can be with and without ruminating.
- *Standing alert:* The preceding behavior is interrupted and the animal raises its head. Characteristic is increased vigilance, testing the surroundings with the three senses: scent, hearing and sight. The disposition for flight is heightened.

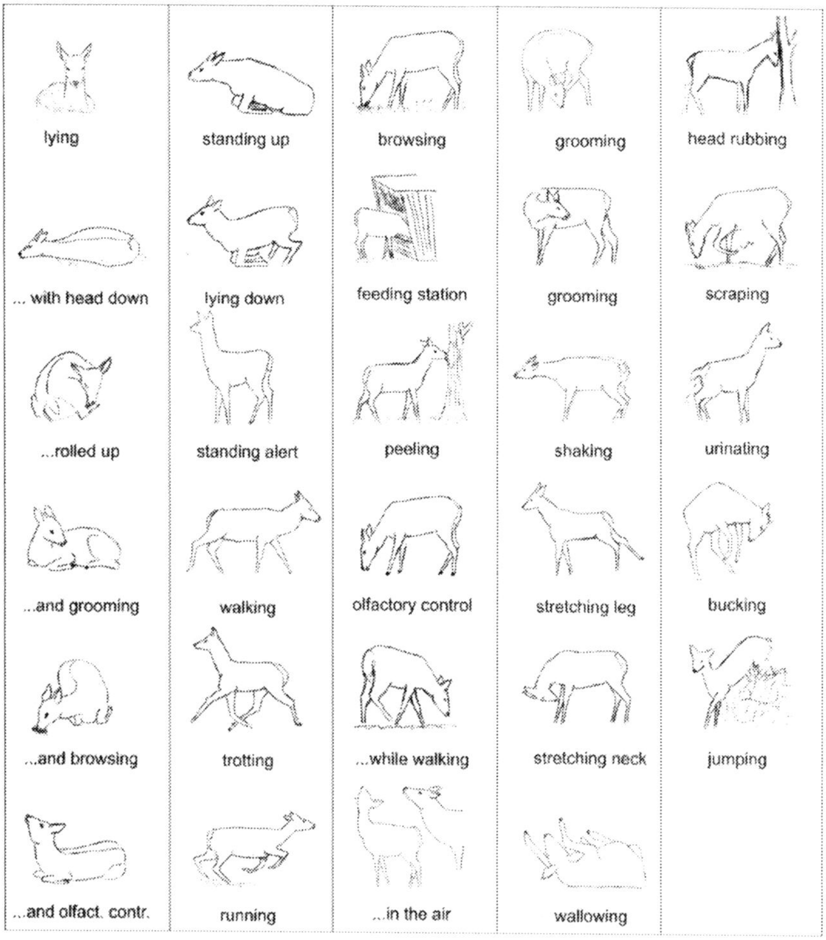

Figure 11. The most common behavioral patterns of roe deer and red deer during the disturbance trials (olfactory control, stretching, wallowing, bucking and jumping were only recorded with red deer). The first column is related to inactive behavior, all other columns to active behavior.

- *Locomotion:* This category includes walking, trotting and running (gallop), but also turning around on the spot and going backwards.
- *Olfactory control:* This behavior is defined as using the sense of smell. Generally the head is lowered, the muzzle moves just above ground level. This behavior pattern can easily be mistaken with foraging, but the mouth stays closed and chewing motion is missing. Also sniffing at higher grasses belongs to this category, in this case

the head is held somewhat higher. This behavior can also be displayed while walking or trotting. When controlling the air the head is tilted upwards, sometimes twisting the neck. Only classified for red deer, since roe deer in comparison rarely showed this behavior type (and was then classified under miscellaneous).

- *Food intake:* While browsing the head is lowered, unless higher grasses and twigs are browsed. Contrary to olfactory control the head movements are jerky due to the plucking of plants. Also chewing movements can be seen. When the head is suddenly raised in alertness plant stems often hang from the jaw. Food intake at the feeding station (hay, supplemental pellets, fodder beets) also belongs to this category.
- *Miscellaneous:* Comfort behavior includes grooming the coat with tongue, teeth, hooves and antlers, furthermore shaking the head or whole body, stretching of neck or legs and wallowing. Other types of behavior that were recorded are: scraping, urinating, head rubbing, bark peeling, drinking, bucking, jumping, combating, lip curling and yawning.

3. Results

First the basic activity patterns of roe and red deer will be presented, afterwards the results of the disturbance trials.

3.1. Basic Activity and Heart Rate Patterns

3.1.1. Roe Deer

3.1.1.1. Circadian and Seasonal Rhythms of Activity and Heart Rate

All individuals had similar activity patterns. The roe deer showed main activity in the morning and evening twilight. In winter the animals were more active in the daytime than at night, in late summer and early autumn (August-mid October) they were mainly inactive in the daytime (Figure 12 and 13).

Daily activity peaks occurred half an hour after sunrise (82.3%) and half an hour after sunset (69.6%). Activity minimum was 1½ hours before sunrise (24.9%) and 2½ hours after sunset (23.4%). In average activity was higher at daytime than at night (Figure 14).

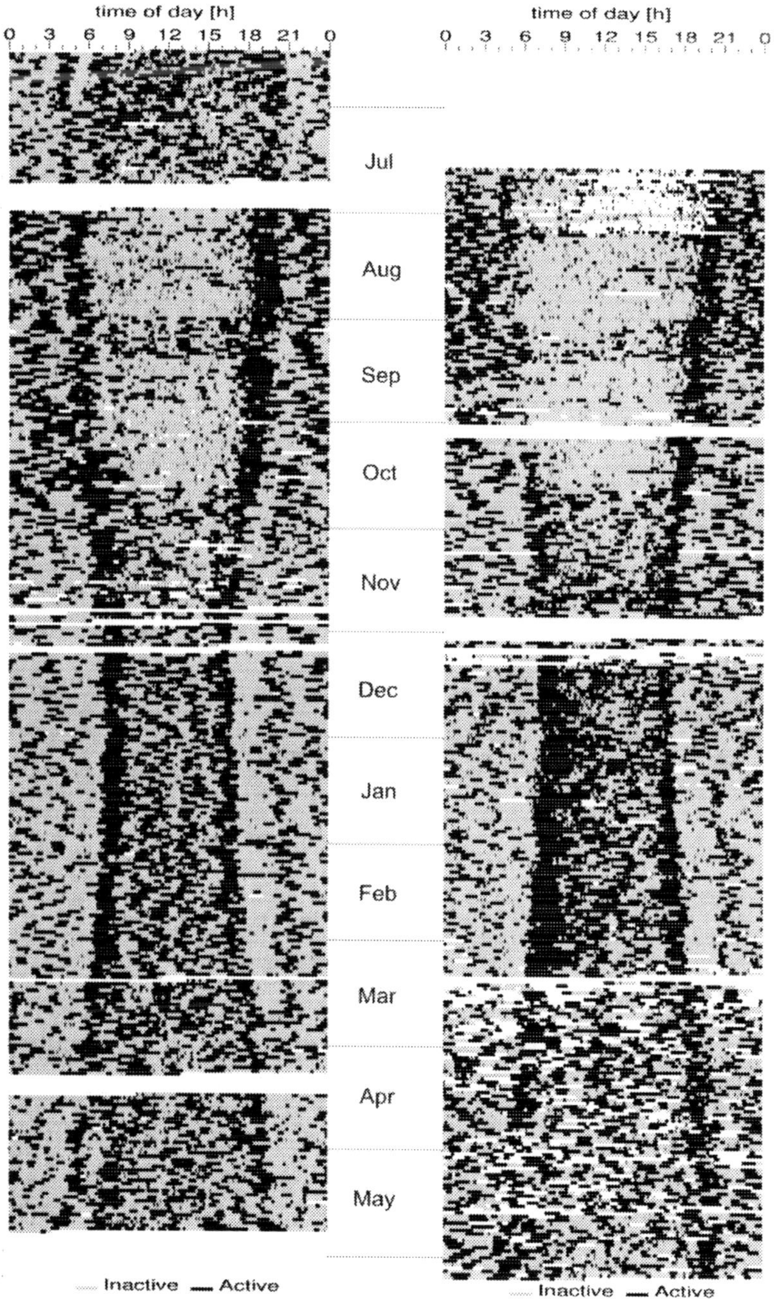

Figure 12. Actograms of two roe deer with a resolution of 1 minute (left: 3-year old male; right: 2-year old female), gray = inactive, black = active, white = missing values.

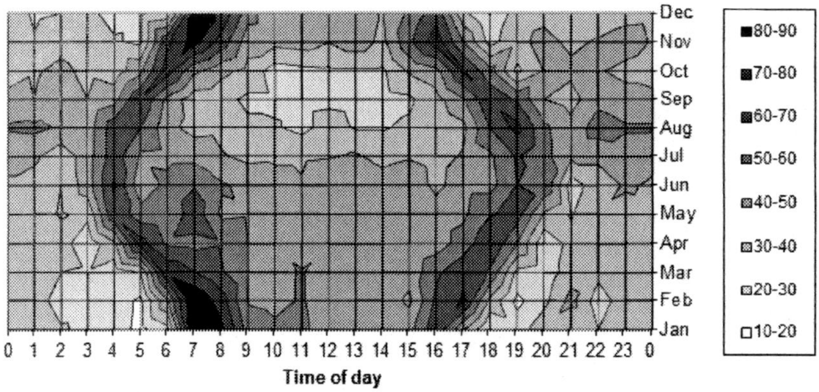

Figure 13. Activity of 14 roe deer (in % of time).

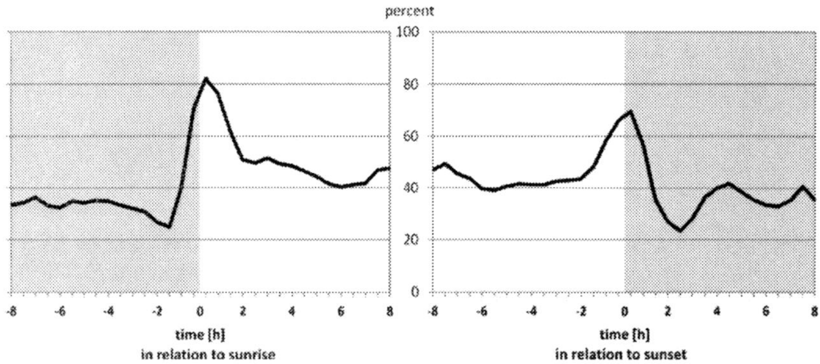

Figure 14. Daily activity of roe deer in relation to sunrise and sunset. Gray shaded areas are night.

The mean monthly activity throughout the year varied greatly among individual roe deer (Figure 15). Roe deer No. 1 (2 year-old female) hardly showed any seasonal fluctuations in activity (39-47%). The animal was mostly kept in a small pen except for September to November when it was transferred to a medium-sized enclosure for disturbance trials. Roe deer No. 3 (3 year-old female) was mainly in a large enclosure and showed great variation in mean monthly activities ranging from 41% in January to 63% in May. The animal was transferred to the medium-sized enclosure for disturbance trials from mid-October until mid-November, which may be the reason for the increase of activity in this time of the year. Male roe deer were in general less active than females (Figure 16). Roe deer No. 22 (3 year-old buck) showed low activity after the rut from August to October (35%) although the buck was involved in the disturbance trials from mid-September to mid-November. Highest activity was in April (49%). Roe buck No. 31 showed little variations ranging from 36% to 45%.

In general the roe deer kept in small pens showed less variation in amount of activity than roe deer in large enclosures (Figure 17).

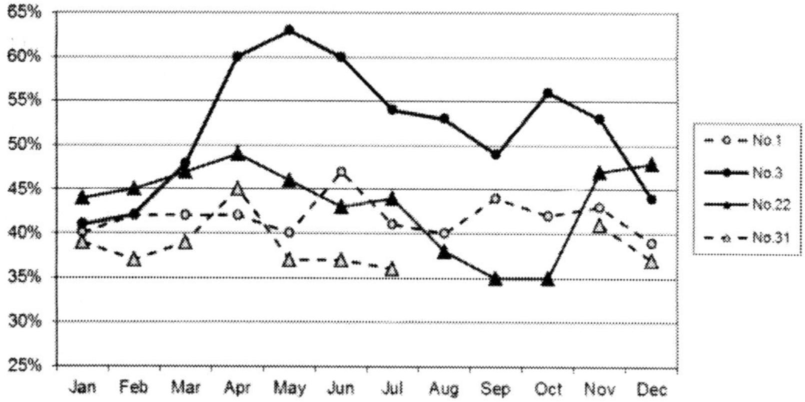

Figure 15. Mean monthly activity (in %) of 4 individual roe deer. (round marks = female, triangle marks = male, dotted line = mainly in small pen, continuous line = mainly in large enclosure). Disturbance trials were carried out from Sep-Nov for Nos. 1 and 22, and Oct-Nov for No.3.

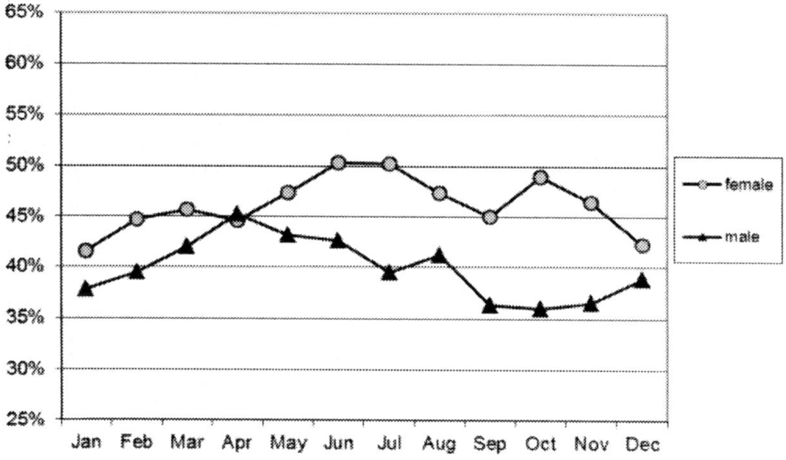

Figure 16. Mean monthly activity (in %) separately for male and female roe deer.

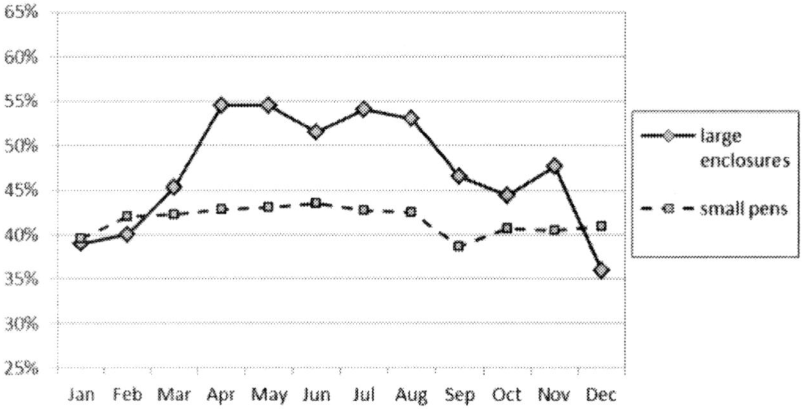

Figure 17. Mean monthly activity (in %) separately for roe deer in large enclosures and in small pens.

Although the seasonal changes in activity varied greatly between individuals, all animals showed very similar circadian patterns. In late summer and early autumn roe deer were less active at daytime compared to the rest of the year. They were more active at night in this time of the year (Figure 13).

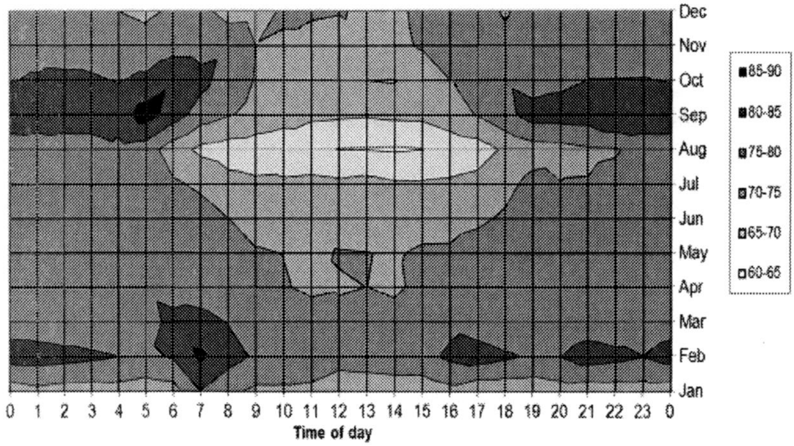

Figure 18. Resting heart rate (beats/minute) of 14 roe deer.

In winter and spring resting heart rate showed very little variations during the day. In late summer and autumn resting heart rate was in contrast much higher at night than at daytime (Figure 18).

3.1.1.2. Influencing Factors on Activity and Heart Rate

A variance analysis was carried out incorporating the input factors daytime, season, sex, size of enclosure, and weekday/weekend (i.e. more/less human activity). All factors had a highly significant influence on activity ($p<0.001$), influence on resting heart rate was significant for all factors except for weekend/weekday. The by far greatest impact was daytime, followed by sex, enclosure type and season.

Roe deer were most active at twilight. In general they were more active during daylight than at night (Table 3).

Table 3. Activity and heart rate of roe deer differentiated by time of day
(twilight = 1 hour before to one hour after sunrise/-set)

	night	twilight	daylight
activity	33.8	64.9	45.9
heart rate	83.8	89.1	81.4
resting heart rate	76.5	76.6	72.4

Altogether does were more active than bucks, yet heart rate of does was lower (Table 4). Both bucks and does were more active during daytime.

Table 4. Activity and heart rate of roe deer differentiated by sex

	male	female	total
total activity	39.7	46.0	43.4
daytime activity	41.7	48.9	45.9
night activity	31.8	35.2	33.8
ratio day/night	1.3	1.4	1.4
heart rate	95.2	75.6	83.9
resting heart rate	85.9	67.0	75.0

Strong interactions exist between daytime and season and between daytime and enclosure type. Roe deer in the large enclosure were in total more active than those in the small pen, yet at daytime activity was higher in the small pens (located next to the research institute with more human activity) (Table 5).

The influence of workdays (with many disturbances) versus weekends (undisturbed) was tested only for daylight time (9 – 15h) when the most disturbances took place (Table 6). Roe deer were more active on weekdays than on weekends (less personal at the research institute and therefore less uncontrolled disturbances). However, the influence on activity was considerably less than the other factors mentioned. Resting heart rate was independent of day in the week (Table 6).

Table 5. Activity and heart rate of roe deer differentiated by enclosure size

	small pen	large enclosure
activity	43.3	47.5
daytime activity	46.4	44.9
night activity	31.2	39.4
ratio day/night	1.5	1.1
heart rate	82.8	83.7
resting heart rate	73.0	72.7

Table 6. Activity and heart rate of roe deer differentiated by weekday/ weekend

	weekday	weekend
day activity (9h-15h)	43.8	38.4
heart rate	82.4	80.3
resting heart rate	73.0	73.1

Total activity was highest during fawning and rut (Table 7). Day activity was highest in spring (establishment of territories) and during fawning and very low post rut. In rut, and especially after rut roe deer were more nocturnal.

Table 7. Activity and heart rate of roe deer differentiated by season
(spring = March/April, fawning period = May/June, rut = July/mid August, post rut = mid August - October, winter = November – February)

	spring	fawning	rut	post rut	winter
activity	43.2	47.4	46.5	41.8	42.5
daytime activity	49.1	51.4	42.7	31.8	48.9
night activity	28.8	32.8	44.2	44.3	30.4
ratio day/night	1.7	1.6	1.0	0.7	1.6
heart rate	78.6	81.6	74.7	86.3	81.8
resting heart rate	69.5	71.4	65.4	76.0	73.8

Particularly for roe bucks during rut and post rut the ratio daytime/nighttime activity was negatively correlated with ambient temperature. For the other seasons correlation was negligible (Table 8, Figure 19).

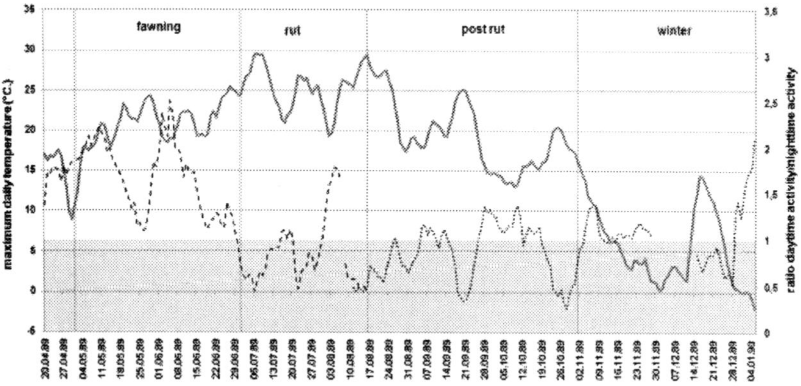

Figure 19. Ratio daytime/nightime activity of two roe deer individuals plotted against maximum daily air temperature. Values are 5-day running means. Gray full line: temperature, dashed line: roe deer No. 27, dotted line: roe deer No. 21. Values under 1 indicate nocturnal, above 1 diurnal behavior.

Activity for the time periods night (21-3), morning (3-9), daytime (9-15) and evening (15-21) was compared for full moon (±2 days) and new moon (±2 days). No significant differences could be found for roe deer in any of the time periods (Mann-Whitney-U-Test).

Table 8. Pearson product-moment correlation coefficient of ratio daytime/nighttime activity of roe deer and daily maximum air temperature, averaged for all individuals

	spring	fawning	rut	post rut	winter
female	-0.07	-0.20	-0.17	-0.29	-0.02
male	0.12	-0.10	-0.60	-0.51	-0.08
total	0.01	-0.15	-0.42	-0.41	-0.06

3.1.1.3. Food Intake

Feeding of roe deer was defined as being active with the head at least 20 sec/min down. This can be only seen as index of feeding, not as absolute feeding time. Feeding patterns are similar to overall activity patterns (Figure 20). Whereas in winter feeding was more evenly spread during daytime and at night, in April and May distinct peaks can be seen at 1:00, 6:00 (dawn), 9:00, 13:00, 15:00, 18:00 (dusk) and 22:00.

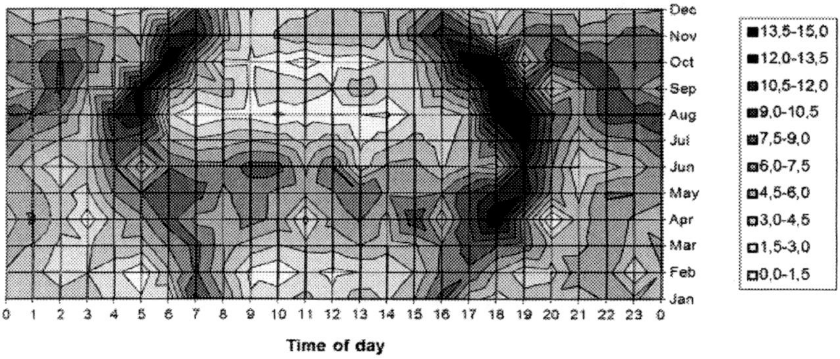

Figure 20. Food intake (in %) of 5 roe deer. Values for March were interpolated from February and April.

3.1.1.4. Sleep

"Sleep" was defined as lying (inactive) with the head on the ground. Sleep could only be discriminated in animals with implanted tilt switch or with repeater system (Table 1 in methods section).

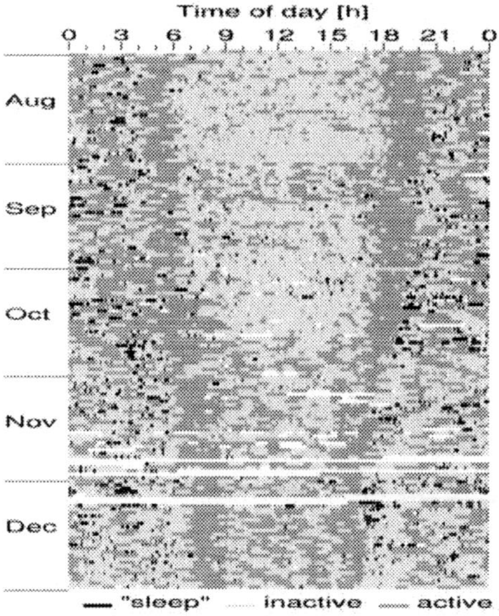

Figure 21. Example of an actogram including sleep phases (2-year old female).

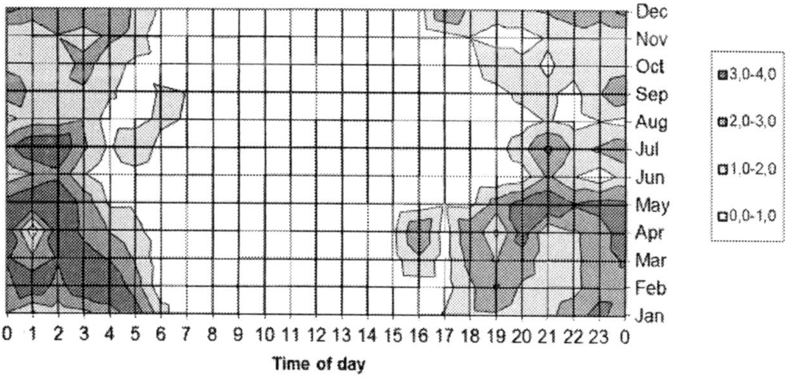

Figure 22. Sleep (in %) of 5 roe deer. Values for March were interpolated from February and April.

Sleep mostly occurred only in the night even in August and September when the deer were mostly inactive in the daytime (Figs. 16 and 17).

3.1.2. Red Deer

3.1.2.1. Circadian and Seasonal Rhythms of Activity and Heart Rate

Just as roe deer the red deer showed strong crepuscular activity. Whereas roe deer had higher activity in the dawn than at dusk, red deer were more active at dusk (Figure 23 and 24). In summer red deer had increased activity in late afternoon, and most of the year also around midnight. Only in the short nights of July there was no increased phase of activity.

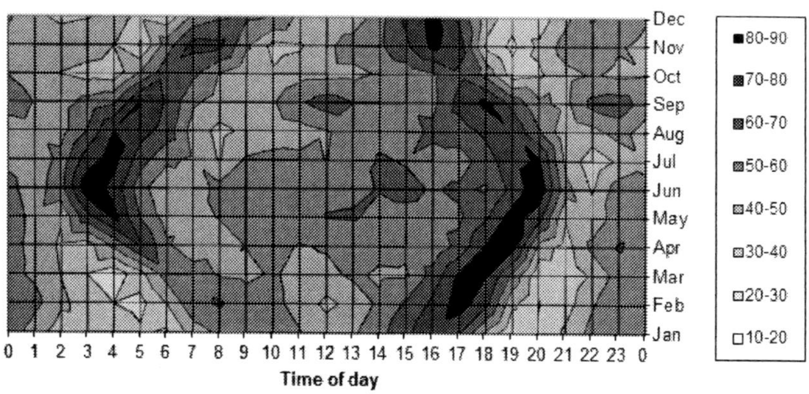

Figure 23. Activity (in %) of 8 red deer.

Daily activity peaks occurred half an hour after sunrise (71.1%) and half an hour after sunset (84.8%). Activity minimum was 2 hours before sunrise (34.9%) and 3 hours after sunset (34.6%). In average activity was higher at daytime than at night (Figure 24).

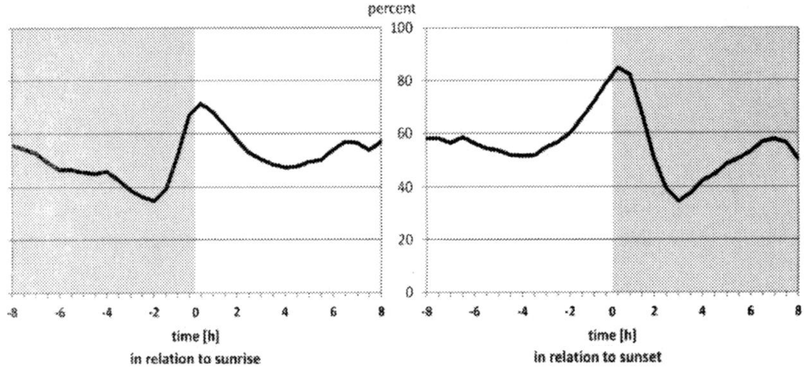

Figure 24. Daily activity of red deer in relation to sunrise and sunset. Gray shaded areas are nighttime.

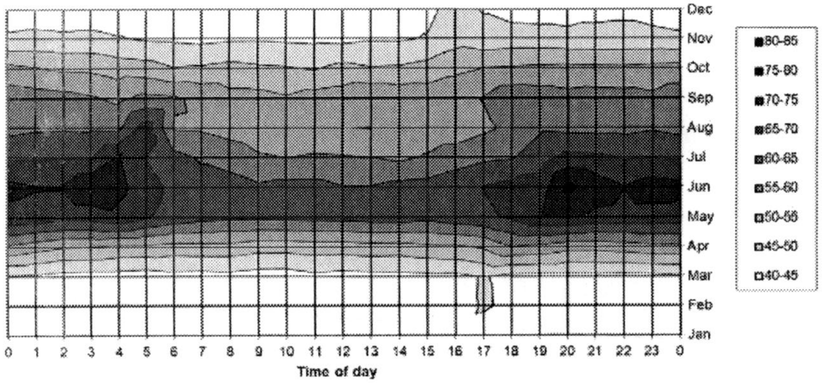

Figure 25. Resting heart rate (beats/minute) of 8 red deer.

Resting heart rate was considerably higher in summer than in winter (Figure 25). Whereas mean resting heart rate was about 44 beats/minute from December to February, resting heart rate averaged 71 beats/minute in June.

3.1.2.2. Influencing Factors on Activity and Heart Rate

Red deer were most active at twilight (Table 9). Mean heart rate was higher at dawn and dusk due to the activity peaks in this time.

Table 9. Activity and heart rate of red deer differentiated by time of day (twilight = 1 hour before to one hour after sunrise/-set)

	night	twilight	daylight
activity	46.4	74.4	54.5
heart rate	59.0	66.8	61.6
resting heart rate	54.1	55.0	55.4

Altogether hinds were more active than stags, especially in the daytime (Table 10). The ratio of daytime activity and night activity was higher for hinds. In contrast to the hinds, stags were more nocturnal.

Table 10. Activity and heart rate of red deer differentiated by sex

	male	female	total
total activity	46.0	56.9	53.8
daytime activity	39.2	60.5	54.5
night activity	47.3	46.0	46.4
ratio day/night	0.8	1.3	1.2
heart rate	61.3	61.6	61.4
resting heart rate	53.7	56.4	54.8

Total activity was highest during calving, summer and rut, and lowest in winter (Table 11). Unlike roe deer the ratio daylight activity to night activity was quite balanced throughout the year.

Table 11. Activity and heart rate of red deer differentiated by season (spring = March/Mid. May, calving period = Mid May/Mid. June, summer = Mid June-August, rut = September-Mid October, post rut = Mid October - November, winter = December – February)

	spring	calving	summer	rut	post rut	winter
activity	52.8	61.6	59.6	58.3	51.3	49.2
daytime activity	53.2	59.5	57.2	57.7	51.6	53.0
night activity	45.0	53.1	50.1	51.7	44.9	42.8
ratio day/night	1.2	1.1	1.1	1.1	1.1	1.2
heart rate	60.6	80.1	73.3	65.9	55.3	48.2
resting heart rate	54.2	70.7	64.6	56.7	48.0	43.5

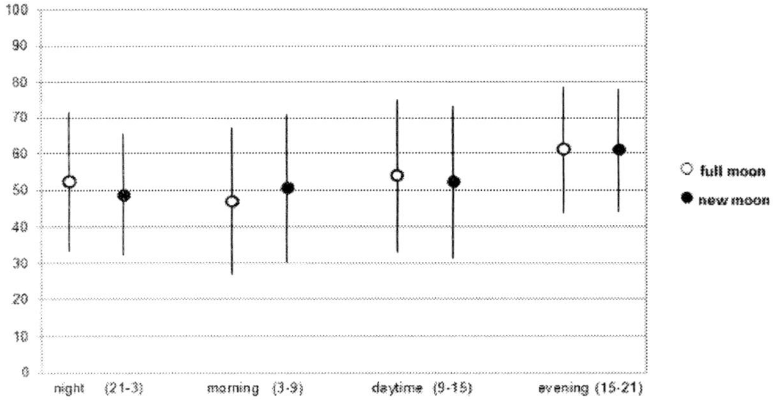

Figure 26. Mean activity in % of red deer at full moon ±2 days and new moon ±2 days for four time periods (n=511 days full moon, 572 days new moon). Bars indicate standard deviation. Differences for night and morning are significant.

Activity for the time periods night (21-3), morning (3-9), daytime (9-15) and evening (15-21) was compared for full moon (±2 days) and new moon (±2 days). In full moon nights red deer were more active (Mann-Whitney-U-Test, $p<0.01$) than in new moon nights and in the morning of full moon nights red deer were less active ($p< 0.001$). No significant differences could be found for daytime and evening hours (Figure 26).

3.1.2.3. Food Intake

Feeding of red deer was defined as being active with the head at least 40 sec/min down. Feeding patterns are similar to overall activity patterns. Highest feeding activity was mainly in the evening (Figure 27).

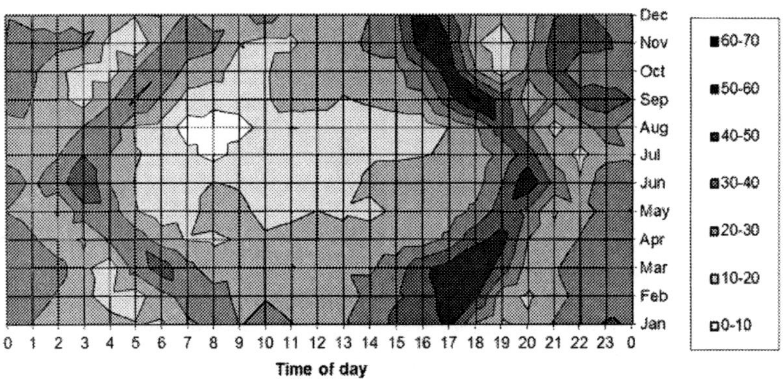

Figure 27. Feeding (in %) of 8 red deer.

3.1.2.4. Sleep

Sleep (lying with head on the ground) covered only a few percent of the daily time budget (highest in October and lowest in May, Figure 28).

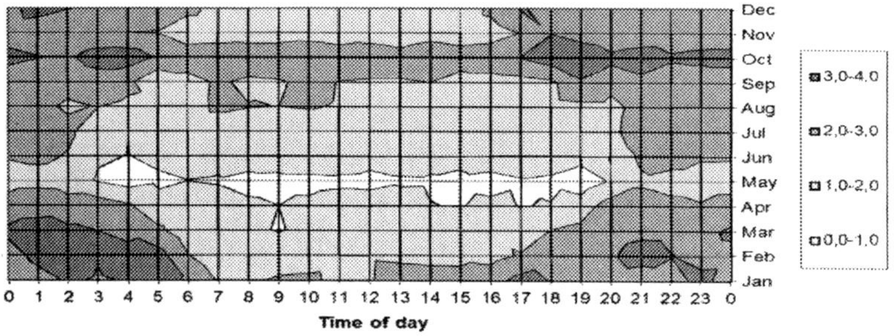

Figure 28. Sleep (in %) of 8 red deer.

3.2. Disturbance Trials

3.2.1. Roe Deer

Since the roe deer showed an abrupt change in general activity in autumn, first being mainly nocturnal and suddenly changing to diurnal behavior on Oct. 21th, the disturbance trials were interpreted separately for both time periods.

3.2.1.1. Choice of Bedding Site

The resting sites of the animals were located in the video recordings. With one exception the three roe deer always bedded in the unmown grass patch from Sept.20th until Oct. 24th noon (Figure 29). After this they rarely lay down in this field, but rather preferred either the small spruce stand or the northeast corner of the enclosure as resting site (Figure 29).

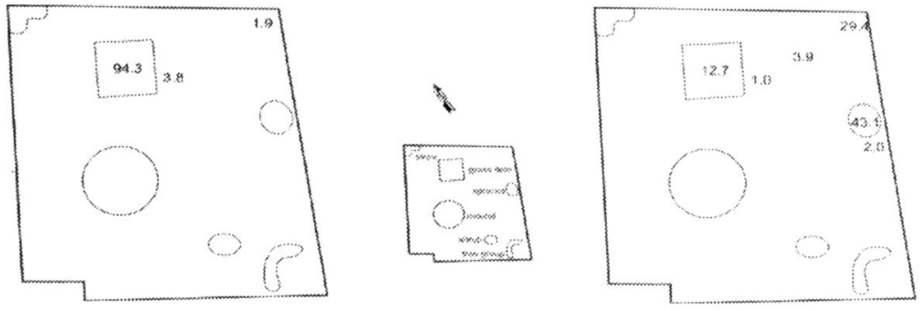

Figure 29. Distribution of roe bedding sites in percentage. Left: Sept.10th to Oct.24th, noon (n = 53); right: Oct.24th, noon to Nov.18th (n=102).

In early autumn the roe deer always bedded in the large grass patch and didn't change this place after haven been disturbed. In late autumn in contrast they chose a new bedding place in 35.3% of the cases when they were disturbed.

3.2.1.2. Activity

Analyses of roe deer were done separately for early autumn (Sept.10 - Oct.20) and late autumn (Oct.21- Nov.18).

Circadian Rhythm

No difference was found in daily activity rhythms between days with disturbance trials (Figure 30) and days without (Figure 31) (Kruskal-Wallis-Test).

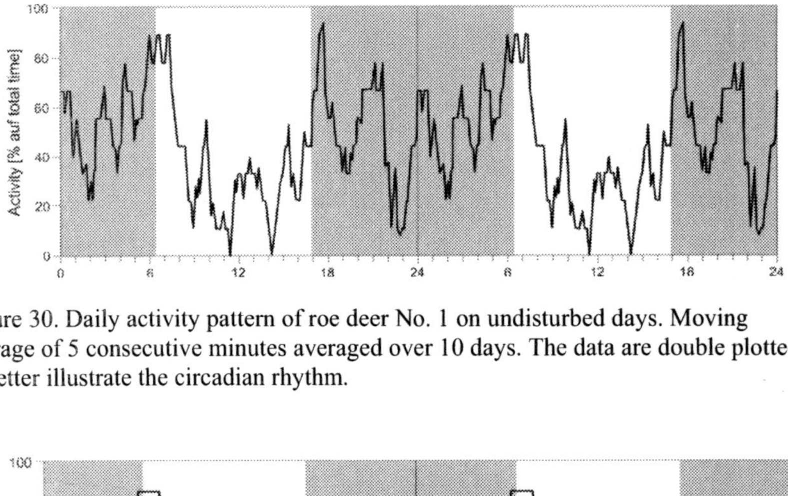

Figure 30. Daily activity pattern of roe deer No. 1 on undisturbed days. Moving average of 5 consecutive minutes averaged over 10 days. The data are double plotted to better illustrate the circadian rhythm.

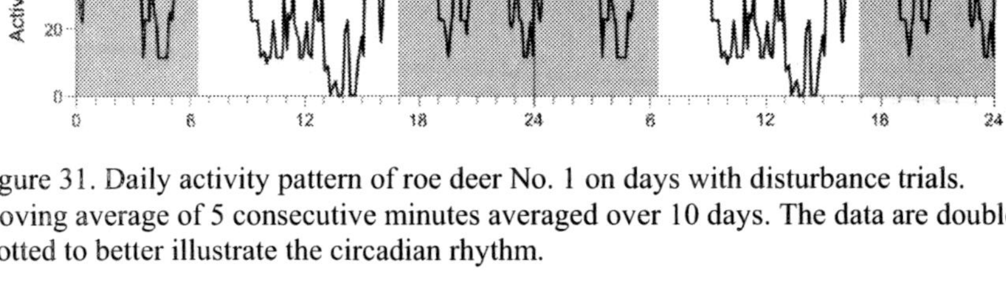

Figure 31. Daily activity pattern of roe deer No. 1 on days with disturbance trials. Moving average of 5 consecutive minutes averaged over 10 days. The data are double plotted to better illustrate the circadian rhythm.

Total Activity

Both roe deer Nos. 1 and 21 were mainly nocturnal in early autumn, independent if disturbance trials were held on that day or not (Tables 12 and 13). The roe buck was less active than the doe (Mann-Whitney-U-Test, $p < 0.05$), both at daytime and in 24 hours. From October 21st on both animals showed an increase in daytime activity ($p < 0.05$). Also roe deer No. 3, which was brought into the enclosure on October 25th was more active at daytime. Contrary to the roe buck ($p < 0.001$) roe deer No. 1 did not change its total amount of activity from early to late autumn. Roe deer No. 3 was more active than the other two deer ($p < 0.01$).

Table 12. Share of time roe deer were active during 24 hours

(No. = animal number, n= number of days, med = median, $\bar{x}$ = arithmetic mean, s = standard deviation, min-max = extrema)

Season	No.	Days without disturbance trials n	med	$\bar{x}$	s	min-max	Days with disturbance trials n	med	$\bar{x}$	s	min-max
Early autumn	1	12	42	40	7.5	39.2-53.2	15	40	41	5.9	32.5-52.6
	22	9	29	28	4.6	19.9-35.9	7	28	28	4.5	27.0-36.5
Late autumn	1	7	44	42	4.3	33.8-47.4	14	42	43	5.2	33.9-53.0
	22	6	37	41	5.9	35.6-50.0	17	43	44	7.0	31.8-62.9
	3	4	53	52	3.5	46.5-54.1	15	55	55	3.9	54.2-60.8

Table 13. Share of time roe deer were active during daytime

(No. = animal number, n= number of days, med = median, $\bar{x}$ = arithmetic mean, s = standard deviation, min-max = extrema)

Season	No.	Days without disturbance trials n	med	$\bar{x}$	s	min-max	Days with disturbance trials n	med	$\bar{x}$	s	min-max
Early autumn	1	12	26	30	14.5	11.2-60.5	15	28	28	8.8	14.7-50.1
	22	9	18	17	5.9	0.5-26.8	7	12	14	6.3	7.7-24.7
Late autumn	1	7	53	50	8.3	35.2-59.5	14	48	50	7.7	63.5-37.6
	22	6	44	45	7.4	36.0-53.3	17	51	50	8.0	31.2-62.0
	3	4	58	56	9.5	46.3-64.8	15	64	62	7.5	51.9-71.7

Both the total activity as well as the amount of daytime activity was not altered by disturbance trials in any of the three roe deer, neither in early nor in late autumn (Tables 12 and 13).

Number and Length of Resting Phases

In early autumn the roe deer had only slightly shorter resting phases on days with disturbance trials (Table 14), whereas in late autumn the mean duration of resting phases was considerably shorter (Kruskal-Wallis-Test).

Table 14. Duration of diurnal resting phases of roe deer (in minutes) on days without, with few (2-4) and with many (5-10) disturbance trials

(No. = animal number, n= number of resting phases, $\bar{x}$ = **arithmetic** mean, s = standard deviation, min-max = extrema)

		no disturbance trials				few disturbance trials				many disturbance trials			
Season	No.	n	$\bar{x}$	s	min-max	n	$\bar{x}$	s	min-max	n	$\bar{x}$	s	min-max
Early autumn	1	21	86	65	4-276	33	70	51	3-202	69	65	47	4-207
	22	35	108	114	6-408	11	133	122	10-415	53	85	88	3-367
Late autumn	1	25	78	41	5-151	19	51	39	7-108	63	43	39	1-155
	22	31	87	52	6-167	46	43	35	5-138	104	33	27	2-108
	3	14	56	37	15-139	15	39	39	4-110	63	31	27	3-91

Table 15. Number of diurnal resting phases of roe deer on days without, with few (2-4) and with many (5-10) disturbance trials

(No. = animal number, n= number of days, $\bar{x}$ = arithmetic mean, s = standard deviation, min-max = extrema)

		no disturbance trials				few disturbance trials				many disturbance trials			
Season	No.	n	$\bar{x}$	s	min-max	n	$\bar{x}$	s	min-max	n	$\bar{x}$	s	min-max
Early autumn	1	7	**6.0**	2.8	3-9	4	**8.3**	1.3	7-10	7	**8.0**	2.0	6-11
	22	9	**7.6**	3.4	3-18	2	**6.0**	0	6	7	**7.6**	2.0	5-11
Late autumn	1	6	**4.5**	0.8	4-6	3	**6.3**	1.5	5-8	7	**8.3**	2.4	6-12
	22	6	**5.3**	2.0	2-8	7	**7.3**	2.2	5-11	11	**9.6**	3.4	6-18
	3	3	**4.7**	1.2	4-6	2	**7.5**	0.7	7-8	8	**7.4**	2.1	6-10

With increasing number of disturbance trials the resting phases shortened. Likewise the number of resting phases at daytime increased with increasing number of disturbance trials. A significant difference (Mann-Whitney-U-Test) could only be found in late autumn (Table 15). The number of resting phases did not differ significantly between early and late autumn on undisturbed days.

Resting Phases during the Course of the Day

Length of resting phases was shorter in the morning and evening hours compared to daytime and at night.

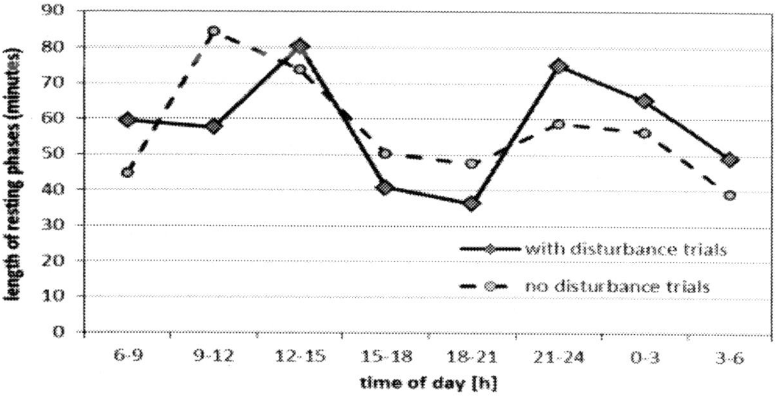

Figure 32. Mean length of resting phases (in minutes) of roe deer during the course of the day on days with and days without disturbance trials in early autumn.

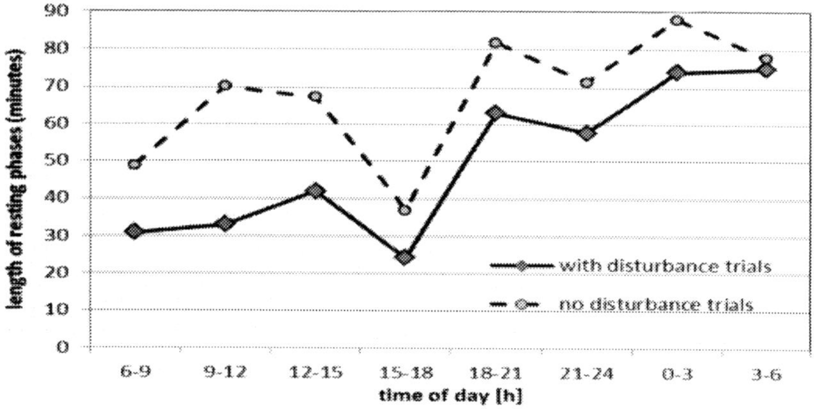

Figure 33. Mean length of resting phases (in minutes) of roe deer during the course of the day on days with and days without disturbance trials in late autumn.

In early autumn no significant differences were found between days with and days without disturbance trials (Figure 32), whereas in late autumn daylight (9-15) resting phases were shorter on days with disturbance trials (Figure 33).

Activation through Disturbance Trials

"Activation" was defined as cases in which an animal`s resting phase is interrupted due to a disturbance and the animal changes over to active behavior. "Activation length" is the length of time an activated animal spends before lying down again. Also with disturbance trials in which video recordings were missing, the activation time could easily be recognized in the transmitted offline data. Disturbance trials in which the animals inevitably had to escape (e.g. the observed animal was lying on the predetermined route of the walking circuit), were omitted from the analysis of activation.

The association of activation among the roe deer was rather low in early autumn (Cole's coefficient of association = +0.38 ± 0.11 between roe deer Nos. 1 and 22). The same applies for late autumn with the coefficients lying in a range between +0.22 ± 0.15 (Nos. 22 and 3) and +0.56 ± 0.18 (Nos. 1 and 3).

Influence of Daytime and Season on Activation of Roe Deer

Both in early and late autumn activation was influenced by time of day (χ^2-Test, $p < 0.05$). The roe deer could be more easily activated in the morning and evening hours than at noon (Figure 34). Roe deer Nos. 1 and 22 were activated more frequently in late autumn (55.4%) than in early autumn (27.3%, χ^2-Test, $p < 0.001$).

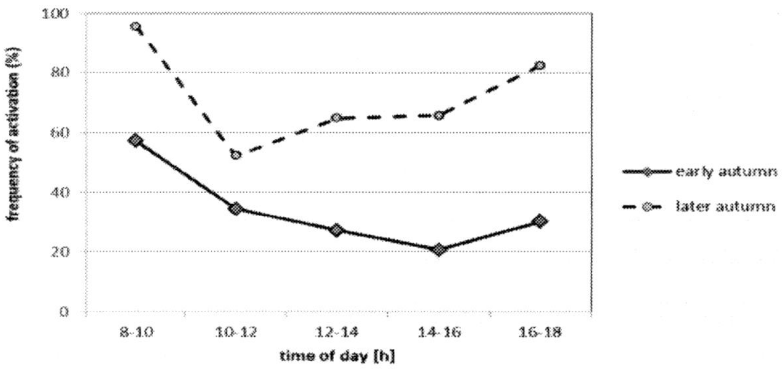

Figure 34. Activation through disturbances in dependence on daytime in early and late autumn. The relative frequency of activation is pooled for two roe deer in early autumn and three roe deer in late autumn.

Dependence of Roe Deer Activation on Type of Disturbance

In early autumn there appears to be a difference of activation frequency by optical (31.4%) and acoustical stimuli (18.8%), yet this seemingly discrepancy was not statistically significant. In late autumn roe deer were significantly more often activated by optical (68.0%) than by acoustical stimuli (21.4%). Furthermore activation by optical stimuli were more frequent in late than in early autumn, whereas acoustical stimuli did not show a significant difference in this time period (χ^2-Test).

Both in early and late autumn the type of disturbance stimuli had different intense effects on activation (χ^2-Test). Generally activation was more easily provoked by the same disturbance type in late than in early autumn (Table 16).

Table 16. Activation (in %) in dependence on disturbance type and season. Numbers of samples are specified in brackets

Disturbance type (*n*)	Early autumn	Late autumn
Round tour of single person (21/6)	19.0	66.7
Round tour of person with dog (11/8)	54.5	100.0
Round tour of group of persons (14/-)	92.9	---*
Round tour with horse(s) (-/6)	---*	66.7
Hang-glider (-/13)	---*	76.9
Tractor (5/10)	20.0	40.0
Ascent of raised blind (12/5)	0.0	40.0
Feeding (23/28)	13.0	71.4
Radio music (9/11)	22.2	9.1
Dog barking (2/11)	0.0	0.0
Chain saw (11/9)	9.1	0.0
Gunshot (17/10)	23.5	80.0

*corresponding trials were not carried out.

Duration of Activation

Individual differences in the length of activation between single roe deer could not be identified, neither in early nor in late autumn (Kruskal-Wallis-Test). In early autumn length of activation was on average 2.7 minutes (median 1 min), in late autumn about 26 minutes (median 16 min.). The difference was highly significant (Kruskal-Wallis-Test, $p < 0.0001$). The duration of activation was independent of the length of the resting phase directly before it. Also there was no relationship between the type of disturbance and the length of the provoked activity phase following.

Yet there was a close connection between time of day and duration of activation (Figure 35).

The lengths of active bouts following activation were longer in morning and late afternoon and shorter at noon in late autumn. But these differences were not statistically significant, probably owing to the small number of samples. The activation initiated by feeding is presumably not responsible for the increase of activation length between 14 and 16h, as the duration of activation caused by feeding and the duration brought about by other disturbance types did not differ significantly for this time period (Mann-Whitney-U-Test).

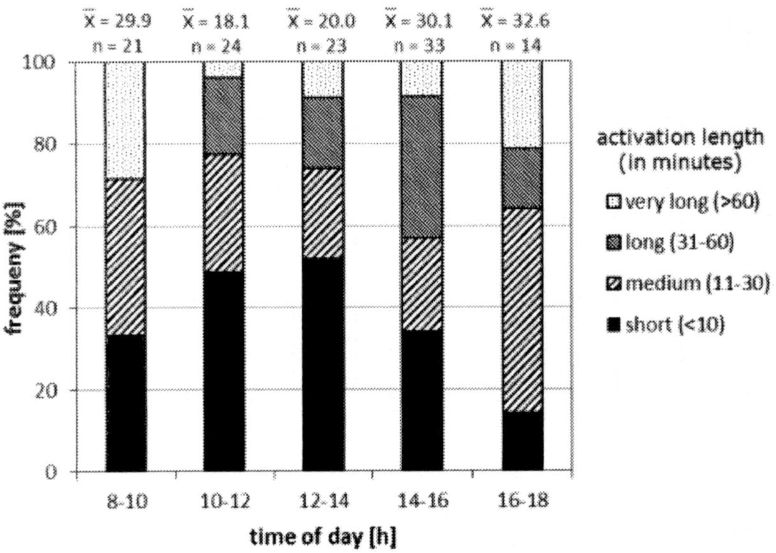

Figure 35. Dependence of activation length on time of day averaged for three roe deer. Above the column the number of samples and the mean duration of activation (in minutes) for each class are specified.

It was also compared to which extent the duration of activation differed between two simultaneously activated roe deer, to draw conclusions on synchronization of behavior. The variation between the length of activation of single roe deer was comparatively high (Table 17).

Mostly the roe buck (No. 22) had lain down at first, in 68.8% of all cases before roe deer No. 1 and in 76.2% of the cases before No. 3. Roe deer No. 3 mostly had lain down as last, only 29.4% of the cases it had lain down before No. 1.

Table 17. Variation in duration of activation (in minutes) between two roe deer

Animal Nos.	n	median	$\overline{x}$	s	extrema
1/22	28	5	17.2	26.4	0-87
1/3	17	14	25.8	29.9	1-89
22/3	21	15	33.0	40.0	1-162

Dependence of Duration of Activation on Type of Disturbance

No significant differences in length of activation of roe deer could be found between varying disturbance stimuli. Generally there was also no difference between duration of activation provoked by optical or acoustical stimuli.

3.2.1.3. Behavior

The behavioral patterns after disturbance trials were dissected in minute entities and compared with the behavior before applying the disturbance stimuli.

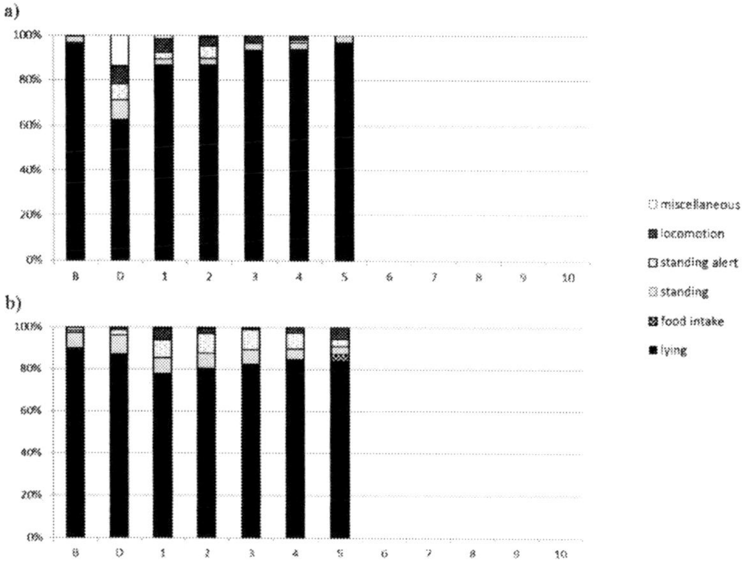

Figure 36. Behavioral pattern of roe deer Nos. 1 and 22 in early autumn (relative share of time for each behavior class) before (B), during (D) and in the single minutes (1,2,3…) after a) optical disturbances trials (504 minutes observation time in total) and b) acoustical disturbances trials (315 minutes observation time in total).

Undisturbed (normal) behavior after disturbance trials returned in the moment when the behavioral pattern did not differ anymore from the pattern before the disturbance. The relative share of time for each behavioral class was calculated for each minute after the disturbance and opposed to the corresponding values of the undisturbed behavior before the disturbance. 250 disturbance trials were analysed. For the time period before disturbance trials 1065 minutes were evaluated, for the period after trials 1142 minutes were drawn upon. In early autumn the behavior of roe deer was only marginally affected (Figure 36).

In late autumn the behavior was affected more intensely and longer than in early autumn (Figure 37). After being disturbed the deer lay less and expended more time in standing alert. After acoustical disturbances roe deer showed only minor changes in their behavioral pattern.

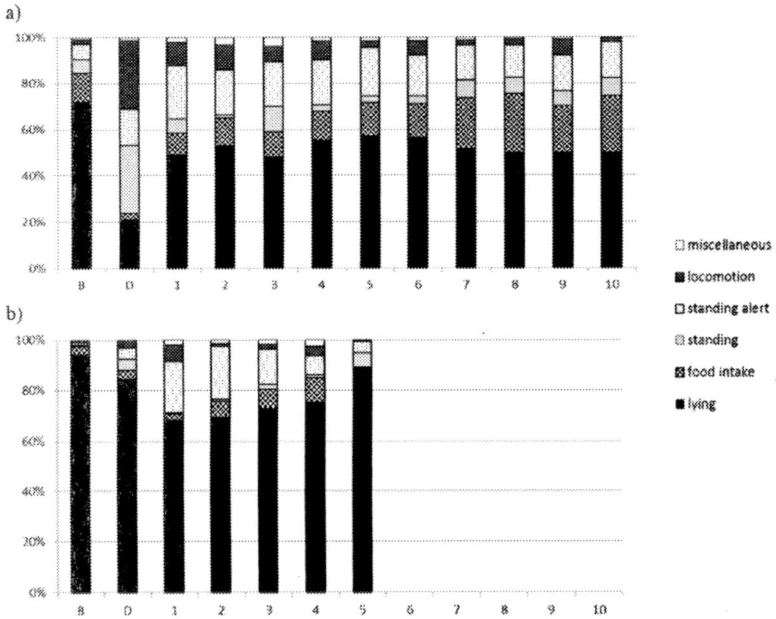

Figure 37. Behavioral pattern of three roe deer in late autumn (relative share of time for each behavior class) before (B), during (D) and in the single minutes (1,2,3…) after a) optical disturbances trials (1275 minutes observation time in total) and b) acoustical disturbances trials (317 minutes observation time in total).

To assess the length of disturbance effect in an animal that was "activated" but did not lie down after the disturbance trials, the behavior of the afterwards active animal was compared with the undisturbed activity of the animal when not exposed to the disturbance stimulus. In the few cases in early autumn, when activated animals did not lie down again, they reacted with increased locomotion and alert standing (Figure 38).

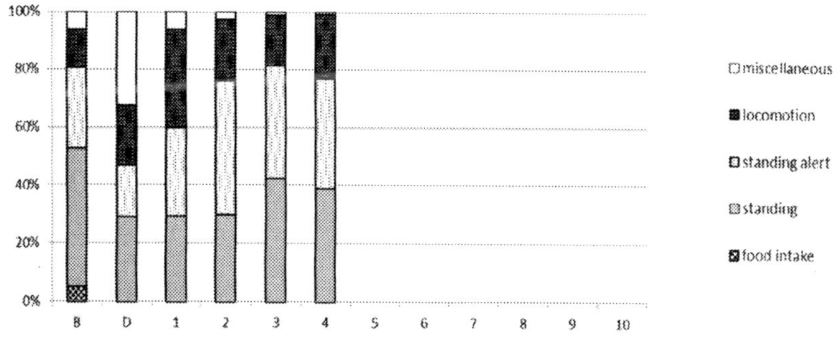

Figure 38. Behavioral pattern of active roe deer in early autumn (relative share of time for each behavior class) before (B), during (D) and in the single minutes (1,2,3...) after optical and acoustical disturbances trials (101 minutes observation time in total).

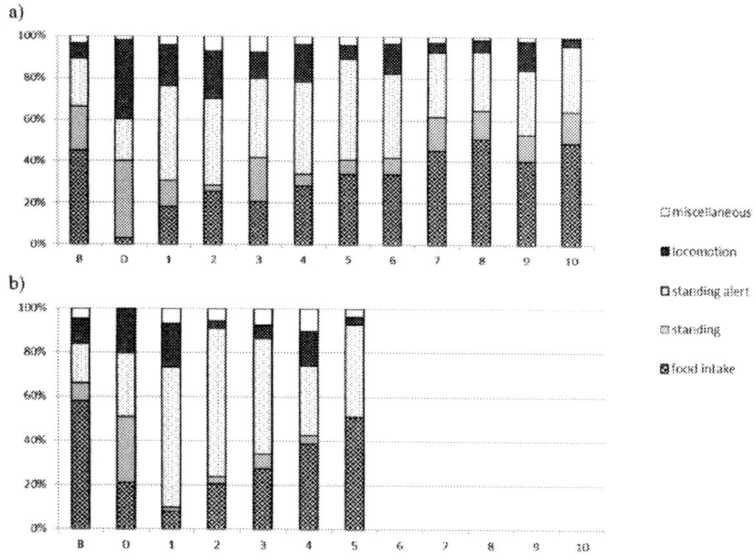

Figure 39. Behavioral pattern of active roe deer in late autumn (relative share of time for each behavior class) before (B), during (D) and in the single minutes (1,2,3...) after a) optical disturbances trials (906 minutes observation time in total) and b) acoustical disturbances trials (49 minutes observation time in total).

In late autumn the food intake after optical disturbance trials was analog to an undisturbed animal after seven minutes (Figure 39a). After acoustical disturbances the feeding activity reached a normal level within five minutes. Yet the time standing alert was still considerably heightened (Figure 39b).

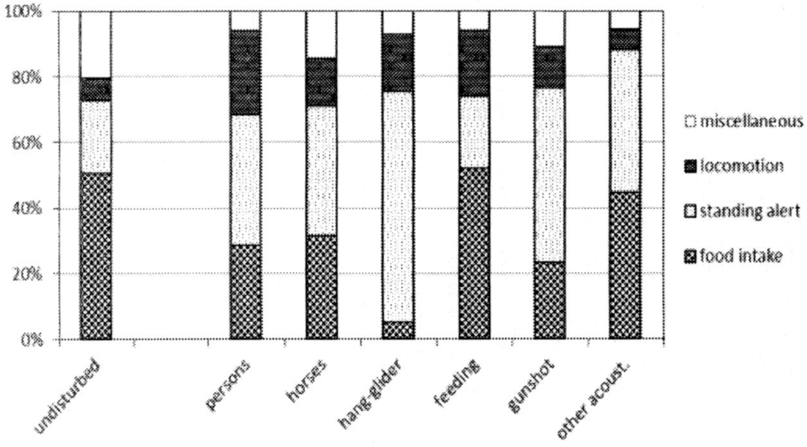

Figure 40. Behavioral pattern of active roe deer (relative share of time for each behavior class) in late autumn before (undisturbed) and after different disturbances trials.

Regarding the varying disturbance stimuli, different behavioral categories were affected in various ways. Locomotion was more pronounced after persons walked by, standing alert after hang-gliders (Figure 40).

3.2.1.4. Heart Rate

Influence of Season on Heart Rate

Daily heart rate was calculated as the arithmetic mean of all heart rate values across 24 hours. The heart rate of the roe buck (No. 22) was on a higher level than the two roe does (Figure 41).

Circadian Rhythm

The mean heart rate of undisturbed roe deer showed periodic changes during the course of the day. It was highest in the morning and lowest in the afternoon (Figure 42). On days with disturbance trials the heart rate showed the same daily pattern as on undisturbed days (Figure 43).

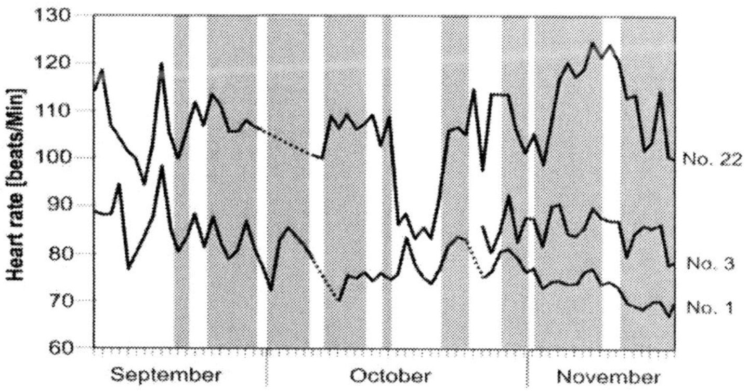

Figure 41. Mean daily heart rate of three roe deer during disturbance trial test period. Days with disturbance trials are highlighted in gray.

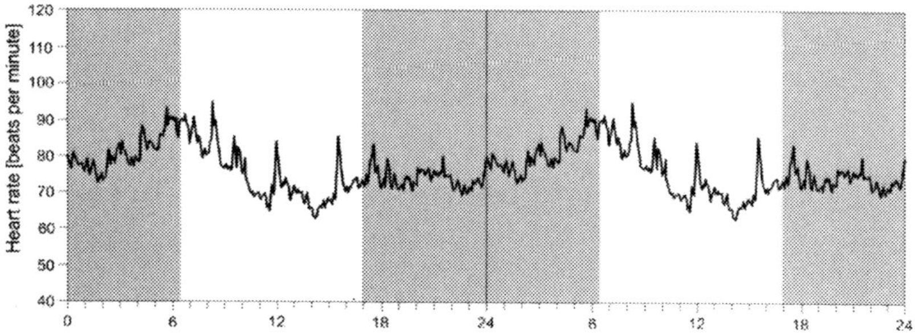

Figure 42. Daily heart rate pattern of roe deer No. 1 on undisturbed days. Moving average of 5 consecutive minutes averaged over 10 days. The data are double plotted to better illustrate the circadian rhythm.

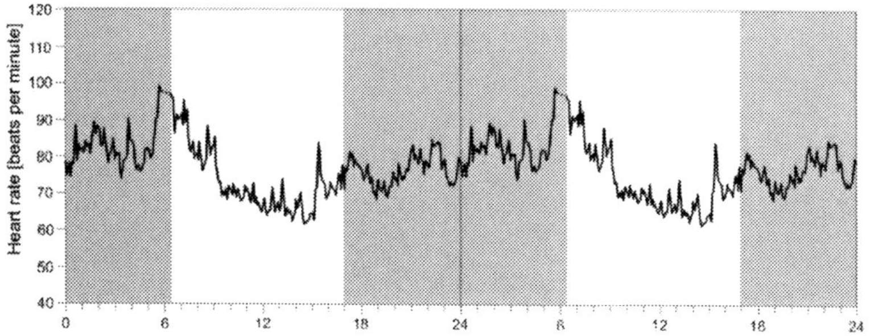

Figure 43. Daily heart rate pattern of roe deer No. 1 on days with disturbance trials. Moving average of 5 consecutive minutes averaged over 10 days. The data are double plotted to better illustrate the circadian rhythm.

Resting Heart Rate

For each resting bout the arithmetic mean of heart rate for each minute was calculated. 1262 resting bouts were taken into account. Kruskal-Wallis-Test was applied to the data. The individual differences in resting heart rate were significant between the three roe deer. Highest heart rate was measured in the roe buck No. 22 (100.8 ± 18.8 beats/min). Doe No. 1 and doe No. 3 had 71.0 ± 6.4 beats/min and 80.3 ± 5.0 beats/min respectively.

Influence of Daytime and Season on Resting Heart Rate of Roe Deer

In early autumn the resting heart rate changed significantly during 24 hours.

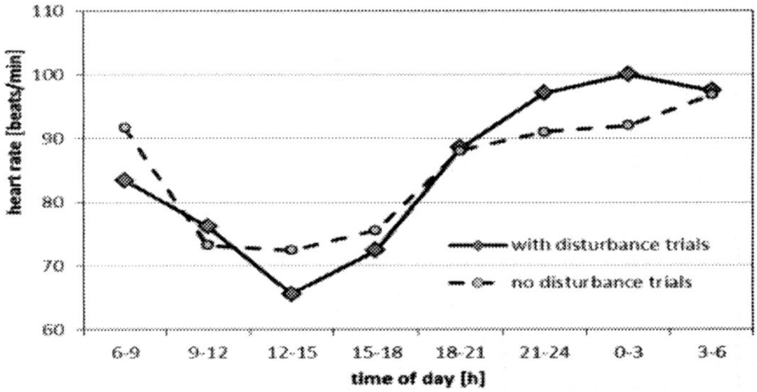

Figure 44. Daily heart rate pattern of roe deer on days with and without disturbance trials in early autumn. 24 hours were subdivided in 8 time segments and the mean heart rate during resting bouts for each fraction calculated.

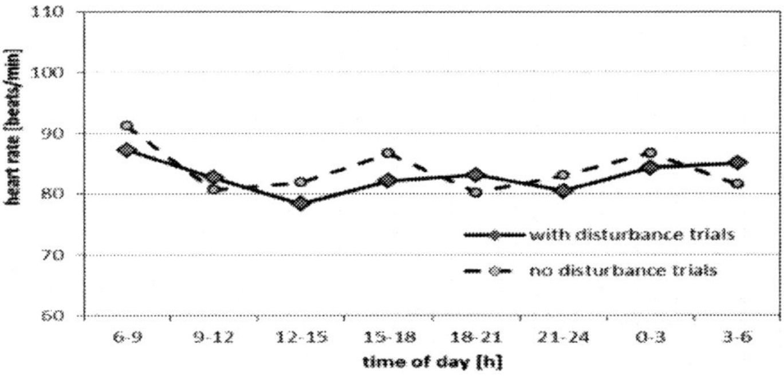

Figure 45. Daily heart rate pattern of roe deer on days with and without disturbance trials in late autumn. 24 hours were subdivided in 8 time segments and the mean heart rate during resting bouts for each fraction calculated.

Resting heart rate was highest in the night and lowest at noon in early afternoon (Figure 44). In late autumn daily fluctuations were not significant (Figure 45). No significant differences could be found in resting heart rate of roe deer between days with and without disturbance trials, neither in early nor in late autumn (Figs. 44 and 45).

Influence of Number of Disturbance Trials Per Day on Resting Heart Rate of Roe Deer

There was no significant relationship between number of disturbance trials carried out per day and resting heart rate of roe deer (Kruskal-Wallis-Test).

Intensity of Disturbance and Increase of Heart Rate

In general resting roe deer showed less heart rate reactions to disturbances than active animals. The magnitude of heart rate reactions towards disturbances increased in late autumn (Tables 18 and 19).

Table 18. Heart rate reactions after disturbances of roe deer in early autumn (2 individuals)

Status of roe deer during disturbance trial	Number of additional heart beats	Duration of increased heart rate (in seconds)	Magnitude of heart rate increase (beats/min)	n
Lying	50	265	11	70
Active	181	378	29	59

Table 19. Heart rate reactions after disturbances of roe deer in late autumn (3 individuals)

Status of roe deer during disturbance trial	Number of additional heart beats	Duration of increased heart rate (in seconds)	Magnitude of heart rate increase (beats/min)	n
Lying	137	263	31	9
Active	316	680	28	95

Influence of Disturbance Type on Heart Rate Reactions of Roe Deer

In early autumn the strongest heart reactions were caused by gunshots. Despite shorter duration of the disturbance effect compared to the other disturbance trials, the strong increase of heart rate caused by gunshots resulted in 119 extra heart beats in the roe buck. Other acoustical disturbances showed the lowest heart rate reactions (Table 20).

In late autumn heart rate reactions were higher than in early autumn. Highest heart rate reactions were caused by persons walking the tour (Table 21).

Table 20. Heart rate reactions after disturbances of roe deer in early autumn separated for different disturbance types and roe deer individuals

	Roe buck (No. 22)				Roe doe (No. 1)			
Disturbance type	Extra heart beats	Duration of heart rate increase	Mean increase (beats/ min)	*n*	Extra heart beats	Duration of heart rate increase	Mean increase (beats/ min)	*n*
Persons	59	246	14	*25*	96	240	24	*29*
Feeding	40	293	8	*11*	19	141	8	*12*
Gunshot	119	110	65	*6*	97	119	49	*10*
Other acoust.	7	186	2	*13*	31	250	8	*11*

Table 21. Heart rate reactions after disturbances of roe deer in late autumn separated for different disturbance types and roe deer individuals

	Roe buck (No. 22)				Roe doe (No. 3)			
Disturbance type	Extra heart beats	Duration of heart rate increase	Mean increase (beats/ min)	*n*	Extra heart beats	Duration of heart rate increase	Mean increase (beats/ min)	*n*
Persons	275	515	32	*8*	288	608	28	*5*
Horse riders	127	380	20	*9*	173	464	22	*6*
Hang-glider	66	197	20	*4*	51	205	15	*5*
Feeding	155	409	23	*6*	176	569	19	*4*
Gunshot	102	257	24	*3*	66	251	16	*4*
Other acoust.	19	152	8	*7*	23	131	10	*4*

Table 22. Heart rate reactions of roe deer in early autumn (2 individuals) after disturbances by persons walking separated for three subgroups

Disturbance type	Number of additional heart beats	Duration of increased heart rate (in seconds)	Magnitude of heart rate increase (beats/min)	*n*
Single person	46	198	14	*26*
Person with dog	80	268	18	*16*
Group of persons	150	306	29	*12*

If disturbance trials with persons walking the round tour are evaluated separately for its subgroups the heart rate reactions are by far higher when a group of persons (at least 3 people) are involved than for single persons

($p<0.01$). A single person with a dog on leash also elicits higher heart rate reactions (Table 22) than a person without dog, but the differences were statistically not significant.

3.2.2. Red Deer

3.2.2.1. Choice of Bedding Site

Red deer did not show special preferences for their bedding sites (Figure 46). Contrary to the roe deer they often lay on the upper end of the enclosure which is nearer to human activities. Often they lay under an elder shrub and on top of the small grass covered mound.

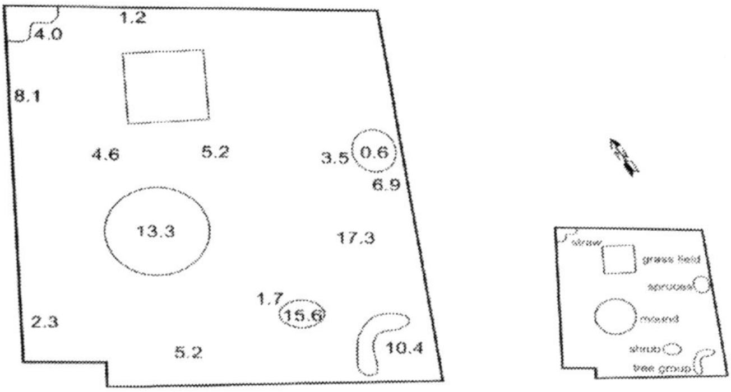

Figure 46. Distribution of red deer bedding sites in percentage (n=173).

Choice of Bedding Site after Disturbances

In 53.6% of all cases in which the animals were activated through disturbance trials a different bedding site was chosen after the trial.

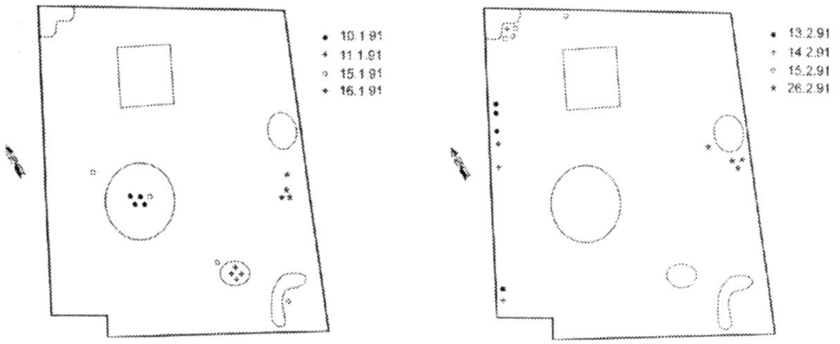

Figure 47. Bedding sites of red deer bedding sites on specific days (each day mapped had at least four disturbance trials).

It is noteworthy that on certain days a favoured bedding site was frequented again for the rest of day even if disturbance trials were carried out, but on the following day a different bedding site was chosen (Figure 47).

3.2.2.2. Activity

Circadian Rhythm

On undisturbed days main activity peaks were in the morning and evening hours with a smaller activity peak in the early afternoon (Figure 48). On days with disturbance trials daytime activity was on a higher level than on undisturbed days. Also nighttime activity showed modifications in rhythm when disturbed at daytime (Figure 49).

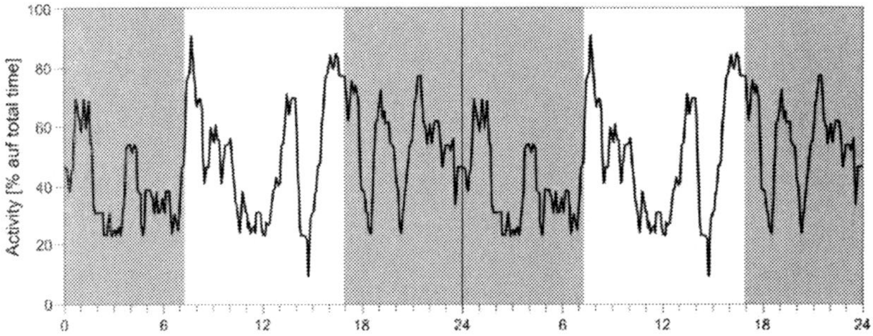

Figure 48. Daily activity pattern of red deer No. 61 on undisturbed days. Moving average of 5 consecutive minutes averaged over 10 days. The data are double plotted to better illustrate the circadian rhythm.

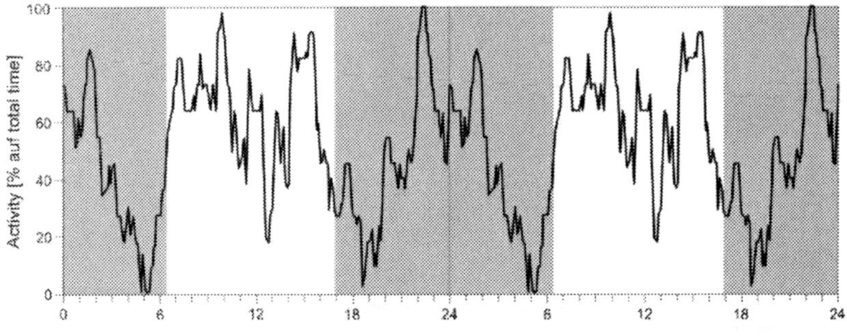

Figure 49. Daily activity pattern of red deer No. 61 on days with disturbance trials. Moving average of 5 consecutive minutes averaged over 10 days. The data are double plotted to better illustrate the circadian rhythm.

Total Activity

On days with disturbance trials the total activity time was slightly increased with the animals spending half an hour less resting per day. The difference was only significant for the calf (No. 51) in January (Table 23). Relating to daylight time the animals reduced the time spent resting by 10 percent (Table 24). The reduction of resting time in 24 hours is effected practically only in the daytime and not compensated for at nighttime.

Table 23. Share of time red deer are active during 24 hours

(No. = animal number, n= number of days, med = median, $\bar{x}$ = arithmetic mean, s = standard deviation, min-max = extrema)

		Days without disturbance trials					Days with disturbance trials				
Month	No.	n	med	$\bar{x}$	s	min-max	n	med	$\bar{x}$	s	min-max
Dec	51	13	48.6	48.2	6.7	32.5-61.1	--	---	---	--	-----
	61	18	51.6	48.7	6.4	35.0-55.6	2	50.6	50.6	11.9	42.2-59.0
Jan	51	9	50.0	49.8	4.1	42.9-56.2	11	55.9	55.1	6.6	44.4-65.5
	61	13	46.4	47.2	4.0	38.6-53.6	18	48.5	49.4	5.9	40.2-63.4
Feb	51	9	53.4	52.9	3.8	46.9-57.7	9	57.5	56.9	7.0	47.3-67.3
	61	14	47.9	47.1	5.0	39.6-55.3	14	50.2	54.7	7.1	45.2-65.6
Mar	51	4	58.1	58.0	1.5	56.1-59.6	4	52.9	61.8	6.9	57.4-72.1
	61	4	54.6	53.5	3.6	48.5-56.5	5	49.7	56.1	7.6	49.2-67.8

Table 24. Share of time red deer are active during daytime

(No. = animal number, n= number of days, med = median, $\bar{x}$ = arithmetic mean, s = standard deviation, min-max = extrema)

		Days without disturbance trials					Days with disturbance trials				
Month	No.	n	med	$\bar{x}$	s	min-max	n	med	$\bar{x}$	s	min-max
Dec	51	14	53.3	52.1	7.4	45.2-62.8	1	59.8	59.8	--	-----
	61	19	56.8	59.1	8.6	39.3-74.4	2	72.5	72.5	2.1	71.0-74.0
Jan	51	9	53.1	52.1	9.5	40.3-69.2	15	68.4	61.2	7.9	54.2-77.7
	61	12	47.1	49.1	7.3	41.7-64.2	18	61.1	62.2	9.7	48.0-84.0
Feb	51	12	56.9	57.0	7.9	50.3-72.3	8	68.5	67.5	9.8	50.6-80.5
	61	12	51.7	51.1	7.6	40.4-66.3	12	62.6	60.7	11.5	43.0-78.9
Mar	51	5	59.0	54.5	4.3	49.8-59.7	4	61.7	60.3	4.5	53.9-63.1
	61	5	55.3	49.3	4.1	54.2-42.8	5	58.2	56.0	5.2	50.0-60.3

Number and Length of Resting Phases

On days with disturbance trials the duration of resting phases was significantly reduced during the daytime (Mann-Whitney-U-Test; Table 25).

Table 25. Duration of diurnal resting phases of roe deer (in minutes) on days without, with few (2-4) and with many (5-10) disturbance trials

(No. = animal number, n= number of resting phases,
$\bar{x}$ = arithmetic mean, s = standard deviation, min-max = extrema)

	no disturbance trials				few disturbance trials				many disturbance trials			
No	n	$\bar{x}$	s	min-max	n	$\bar{x}$	s	min-max	n	$\bar{x}$	s	min-max
51	208	53	37	6-200	33	40	24	10-95	154	37	25	2-122
61	218	62	37	3-174	34	43	29	7-116	151	42	30	5-144

The number of resting phases was the same on days with and without disturbance trials (Table 26).

Table 26. Number of diurnal resting phases of red deer on days without, with few (2-4) and with many (5-10) disturbance trials

(No. = animal number, n= number of days,
$\bar{x}$ = arithmetic mean, s = standard deviation, min-max = extrema)

	no disturbance trials				few disturbance trials				many disturbance trials			
No	n	$\bar{x}$	s	min-max	n	$\bar{x}$	s	min-max	n	$\bar{x}$	s	min-max
51	35	5.0	1.5	2-8	6	4.7	1.6	3-7	17	5.7	2.1	3-10
61	30	4.9	1.6	3-8	8	5.0	1.1	4-7	27	5.4	1.8	1-8

Resting Phases during the Course of the Day

Length of resting phases was shorter during the daytime than at night (Figure 50). On days with disturbance trials resting phases were particularly shorter during daytime (Kruskal-Wallis-Test).

Activation through Disturbance Trials

The association of activation among the two red deer was very high (Cole's coefficient of assosciation = +0.915 ± 0.14). With both animals no seasonal change in activation could be found during the test period (December – March). The red deer calf tended to be more easily activated (76.7%) by optical and acoustical disturbances than the male deer (65.9%), but the difference was not significant (χ^2-Test).

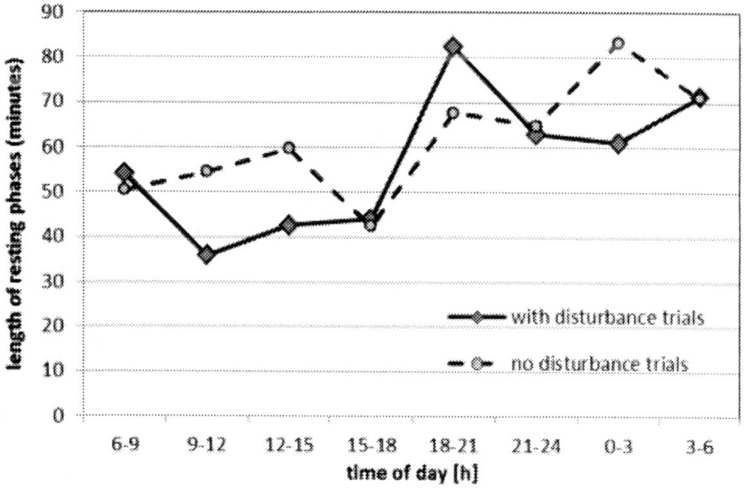

Figure 50. Mean length of resting phases (in minutes) of red deer during the course of the day on days with and days without disturbance trial.

Influence of Daytime on Activation of Red Deer

Both red deer were least activated between 12:00 and 14:00 (Figure 51), although the circadian differences were not significant.

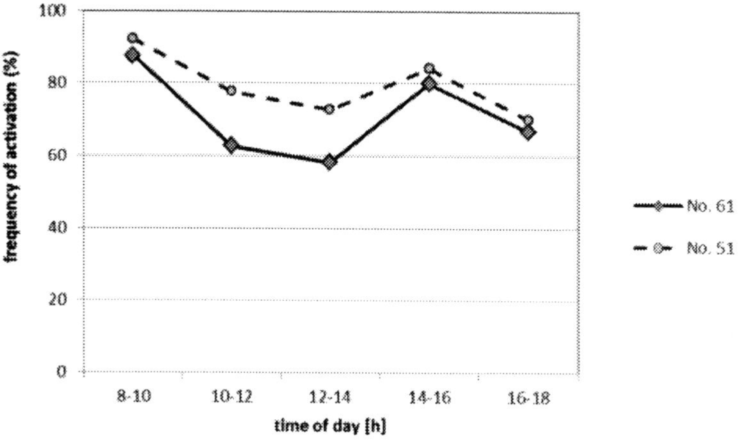

Figure 51. Activation through disturbances in dependence on daytime for red deer nos. 51 and 61.

Dependence of Red Deer Activation on Type of Disturbance

Both red deer were significantly more often activated by optical (87%) than by acoustical stimuli (28%; χ^2-Test, p<0.01). All optical stimuli led to 90-100% activation (Table 27), aside from tractor (28.6%) and hang-glider (33.3%). Of the acoustical stimuli only gunshots caused high activation (53.3%), but this could not be statistically verified.

Table 27. Activation (in %) in dependence on disturbance type and season. (n=number of samples)

Disturbance type	% activated	n
Round tour of single person	100.0	27
Round tour of person with dog	100.0	14
Round tour of group of persons	100.0	14
Round tour with horse(s)	92.0	25
Hang-glider	33.3	12
Tractor	28.6	7
Fire	100.0	10
Feeding	90.9	11
Radio music	12.5	8
Dog barking	25.0	8
Chain saw	9.1	11
Gunshot	53.3	15

Duration of Activation

Length of activation was on average 38.5 minutes. Individual differences in the length of activation between both red deer could not be identified (Kruskal-Wallis-Test).

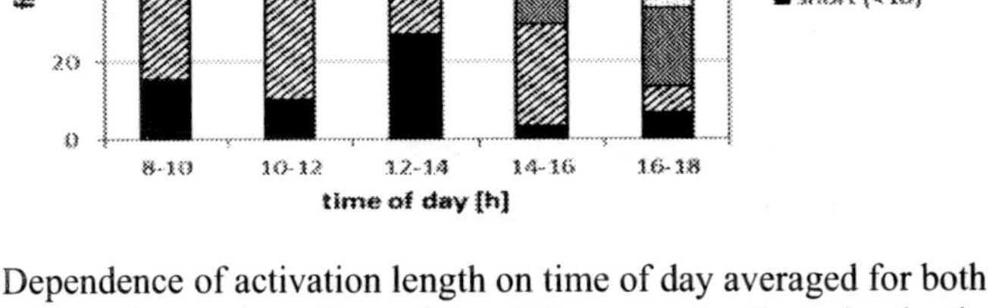

Figure 52. Dependence of activation length on time of day averaged for both red deer. Above the column the number of samples and the mean duration of activation (in minutes) for each class are specified.

There were also no significant differences in length of activation between the months, although there was a trend of decreasing activation length toward end of the winter.

Activation length in dependence of daytime was significant (Kruskal-Wallis-Test). Activation length was longest in the evening (Figure 52)

Differences in length of activation between two simultaneously activated red deer were minimal ($\bar{x}$=10.5 min, median = 5 min, extrema 0-57 min). In equal number of cases one animal was longer active than the other.

Dependence of Duration of Activation on Type of Disturbance

Basically there was also no difference between duration of activation provoked by optical or acoustical stimuli. Yet between individual disturbance stimuli there were very significant differences in length of activation (Table 28; Kruskal-Wallis-Test).

Table 28. Activation length (in minutes) in dependence on disturbance type (n=number of samples)

Disturbance type	median	n
Round tour of single person	34.5	24
Round tour with single person and leashed dog	51.0	15
Round tour of group of persons	55.0	9
Round tour with one horse (walk-gallop)	21.0	24
Round tour with group of horses (walk)	12.0	11
Hang-glider	32.5	4
Tractor	19.0	5
Fire	35.5	6
Feeding	42.0	9
Radio music	3.0	1
Dog barking	18.5	2
Chain saw	3.0	1
Gunshot	68.0	3

3.2.2.3. Behavior

Just as with roe deer the behavioral patterns of red deer before disturbance trials were compared with the time after. Undisturbed behavior returned as soon as the behavior after the trials did not differ significantly from the behavior before the trial. 244 disturbance trials were analysed on basis of seconds. The time span before setting the disturbance stimulus compromised 2484 minutes, after the trial 1869 minutes were charted.

In the first week of the test period red deer were disturbed at least 10 minutes after finishing the disturbance trial with optical stumuli (Figure 53a), however with acoustical disturbance trials there was no significant change to undisturbed behavior after six minutes (Kruskal-Wallis-Test; Figure 53b).

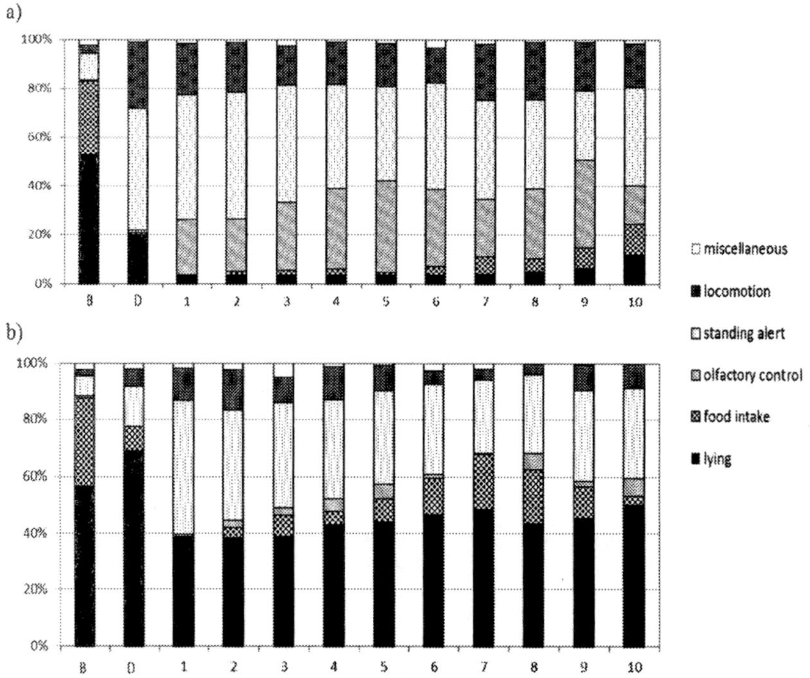

Figure 53. Behavioral pattern of red deer in the first test week (relative share of time for each behavior class) before (B), during (D) and in the single minutes (1,2,3…) after a) optical disturbances trials (642 minutes observation time in total) and b) acoustical disturbances trials (410 minutes observation time in total).

In the last test week the animals showed undisturbed behavior after optical disturbances within 10 minutes (Figure 54a). After acoustical disturbances the behavior was already in a normal range in the following first minute (Kruskal-Wallis-Test; Figure 54b).

When comparing the behavior of animals that stayed active after the disturbance trials with undisturbed active animals, the behavior was still deviant after 10 minutes both for optical and acoustical stimuli (Figure 55)

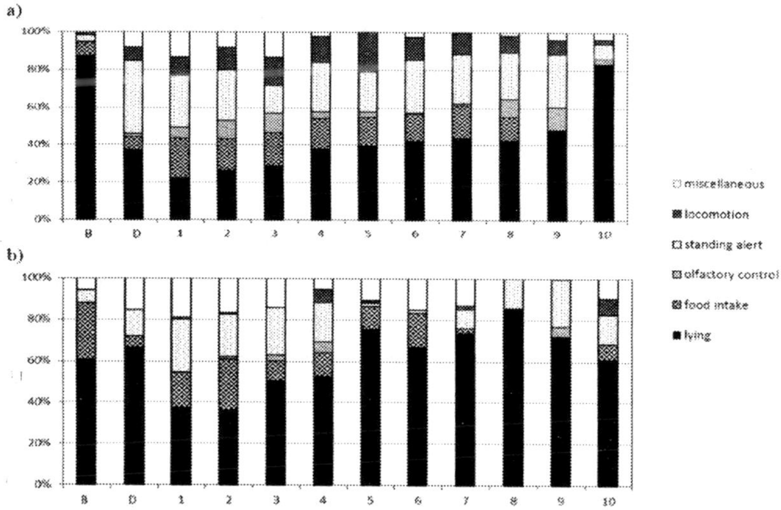

Figure 54. Behavioral pattern of red deer in the last test week (relative share of time for each behavior class) before (B), during (D) and in the single minutes (1,2,3…) after a) optical disturbances trials (476 minutes observation time in total) and b) acoustical disturbances trials (172 minutes observation time in total).

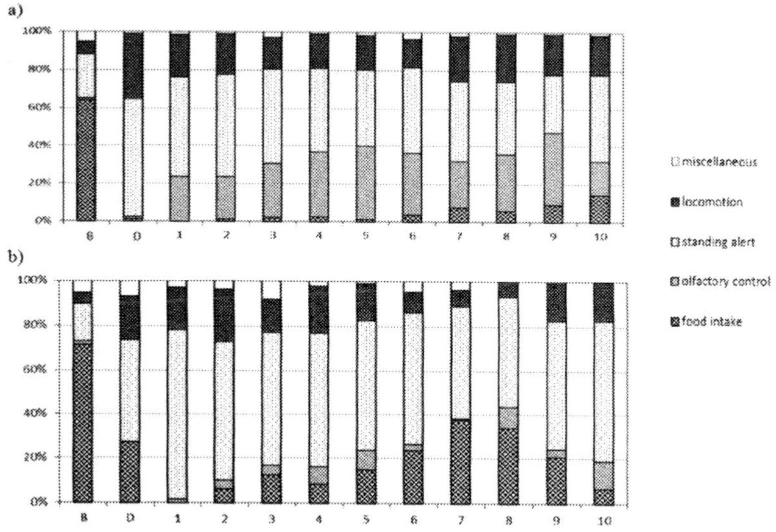

Figure 55. Behavioral pattern of active red deer in the first test week (relative share of time for each behavior class) before (B), during (D) and in the single minutes (1,2,3…) after a) optical disturbances trials (476 minutes observation time in total) and b) acoustical disturbances trials (172 minutes observation time in total).

In the last week active red deer showed at least 8 minutes disturbed behavior after optical stimuli and at least 6 minutes disturbed behavior after acoustical stimuli (Figure 56). The data was not sufficient to evaluate the ninth and tenth or respectively the seventh to tenth minute.

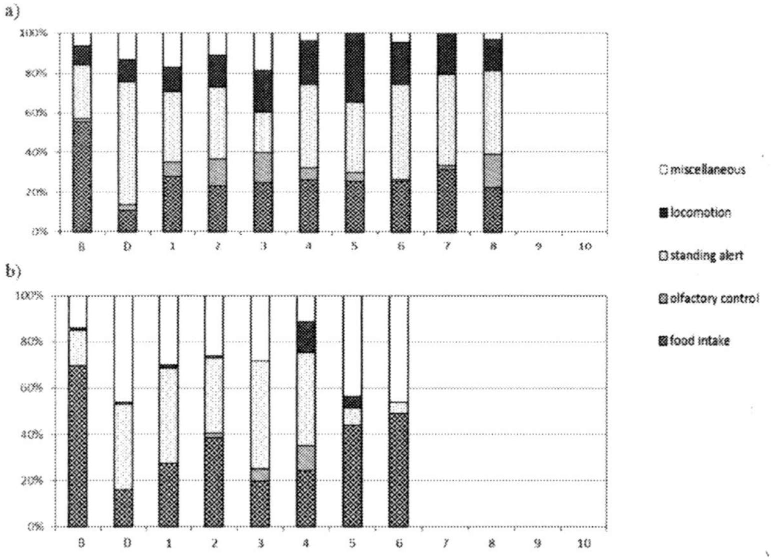

Figure 56. Behavioral pattern of active red deer in the first test week (relative share of time for each behavior class) before (B), during (D) and in the single minutes (1,2,3…) after a) optical disturbances trials (476 minutes observation time in total) and b) acoustical disturbances trials (172 minutes observation time in total).

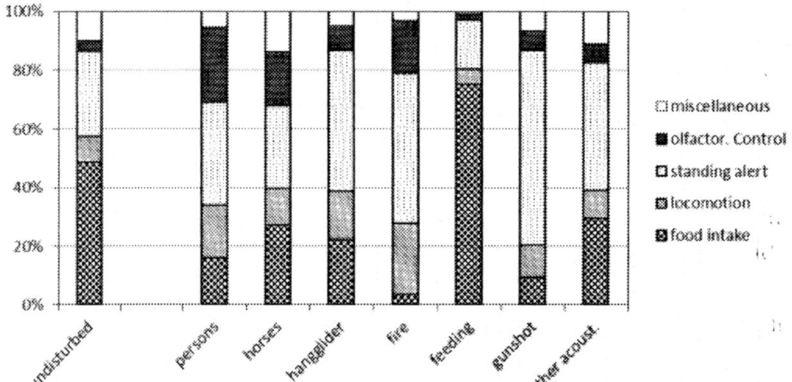

Figure 57. Behavioral pattern of active red deer (relative share of time for each behavior class) before (undisturbed) and after different disturbances trials.

Regarding the varying disturbance stimuli, different behavioral categories were affected in various ways. Olfactory control was most pronounced after persons walked by, standing alert after gunshots. With fire locomotion was more pronounced, after filling the feeding station, food intake increased (Figure 57).

The time span until there was no significant deviation from undisturbed behavior varied with disturbance types. With persons walking in the enclosure and gunshots the behavior still differed significantly after 10 minutes (maximum observation time) in several behavioral classes from undisturbed behavior (Table 29). With other acoustical stimuli, but also with hang-gliders, tractors, and horse riders the behavior was virtually the same after the disturbance as before.

Table 29. Time span (in minutes) after completion of the disturbance trial, in which still significant differences to undisturbed occur in at least one behavioral category

Disturbance trial	Lying	Locomotion	Food intake	Standing alert	Olfact. Contr.	Miscell.
Single person	9	10	4	10	9	---
Person with dog	9	8	9	10	9	---
Group of persons	---	4	---	---	9	---
Horse rider(s)	---	---	---	---	---	---
Hang-glider	---	---	---	1	---	---
Tractor	---	---	---	---	---	---
Gun shot	---	---	10	4	---	---
Other acoust.	---	---	---	---	---	---

3.2.2.4. Heart Rate

Influence of season on heart rate: The mean heart rate decreased continuously from December to mid-March and started to increase afterwards. The heart rate of the male red deer was consistently higher than the female calf deer (Figure 58)

Circadian rhythm: The mean heart rate of undisturbed red deer showed periodic changes during the course of the day. It was highest in the late afternoon (Figure 59). On days with disturbance trials the heart rate was considerably lower, especially between 10:00 and 14:00. At nighttime there was no noticeable change in heart rate between disturbed and undisturbed days (Figure 60).

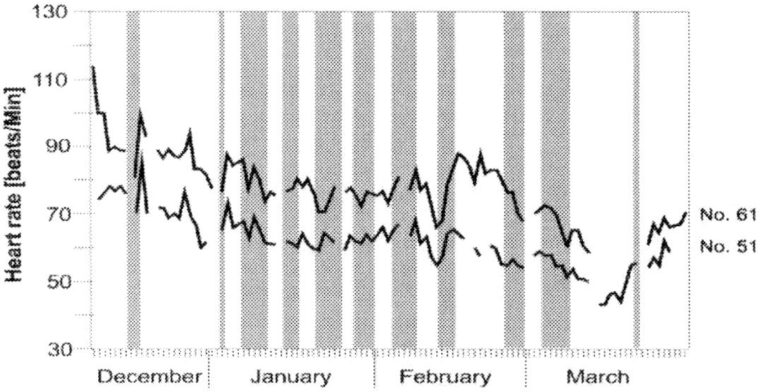

Figure 58. Mean daily heart rate of red deer nos. 51 (female calf) and 61 (male, spike) during the disturbance trial test period. Days with disturbance trials are highlighted in gray.

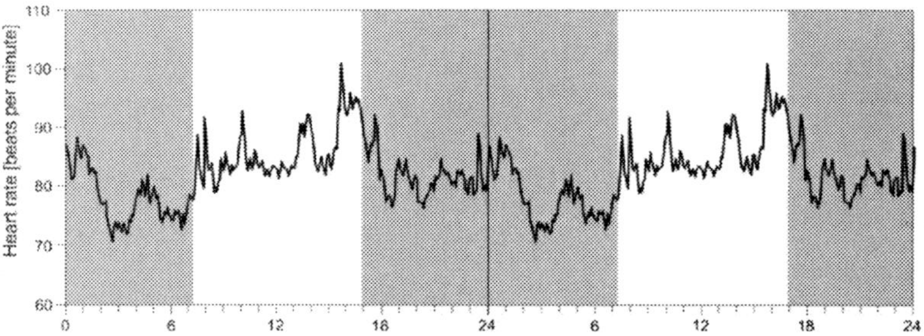

Figure 59. Daily heart rate pattern of red deer No. 61 on undisturbed days. Moving average of 5 consecutive minutes averaged over 10 days. The data are double plotted to better illustrate the circadian rhythm.

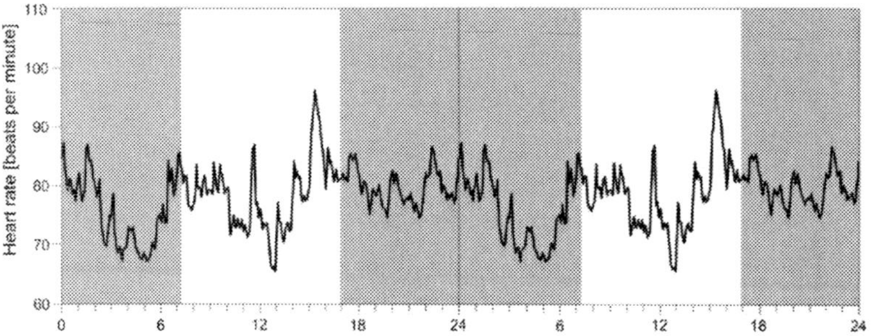

Figure 60. Daily heart rate pattern of red deer No. 61 on day with disturbance trial. Moving average of 5 consecutive minutes averaged over 10 days. The data are double plotted to better illustrate the circadian rhythm.

Resting heart rate: Resting heart rate was calculated as for roe deer, with 1262 resting bouts taken into account. The data was testet with Kruskal-Wallis. The resting heart rate of the male deer (No. 61) was significantly higher (73.9 ± 10.7) than the female calf (58.1 ± 7.6).

Influence of daytime and season on resting heart rate of red deer: Resting heart rate decreased gradually from December (73.2 ± 11.6) over January and February (63.4 ± 10.7) until March (53.7 ± 8.1). Resting heart rate was lower at night and in the morning and highest in the evening (Figure 61). On undisturbed days these fluctuations were not significant in contrast to days with disturbance trials. Generally the circadian variations had a similar pattern for disturbed and undisturbed days, yet the resting heart rate was on a lower level when disturbance trials were carried out (Figure 61).

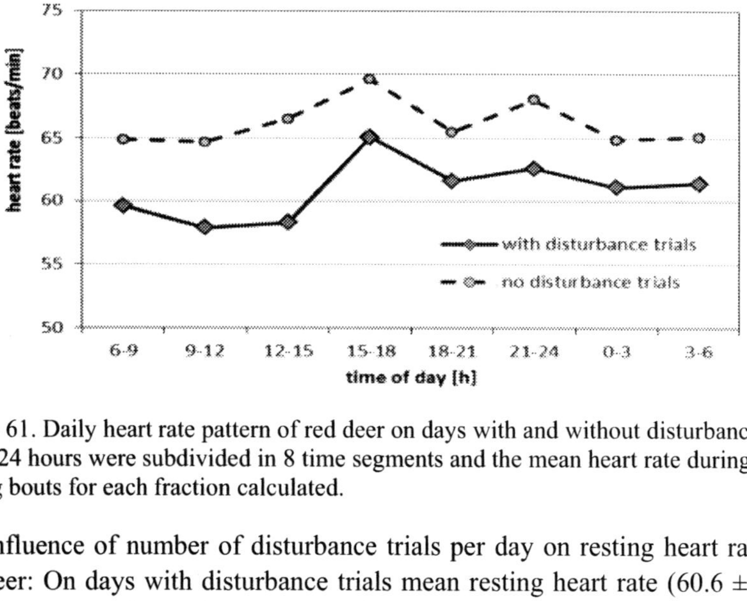

Figure 61. Daily heart rate pattern of red deer on days with and without disturbance trials. 24 hours were subdivided in 8 time segments and the mean heart rate during resting bouts for each fraction calculated.

Influence of number of disturbance trials per day on resting heart rate of roe deer: On days with disturbance trials mean resting heart rate (60.6 ± 9.1) was significantly lower than on days without disturbances (66.0 ± 12.2).

Table 30. Heart rate reactions after disturbances of red deer (2 individuals)

Status of red deer during disturbance trial	Number of additional heart beats	Duration of increased heart rate (in seconds)	Magnitude of heart rate increase (beats/min)	n
Lying	9	102	5	25
Active	257	760	20	201

Increase of heart rate and intensity of disturbance: Resting red deer showed less heart rate reactions to disturbances than active animals (Table 30).

Influence of disturbance type on heart rate reactions of red deer: The greatest heart rate reactions were caused by feeding, especially for the female calf (Table 31). Acoustical stimuli and the hang-glider caused only weak reactions.

Table 31. Heart rate reactions after disturbances of red deer separated for different disturbance types and individuals

	female (No. 51)				male (No. 61)			
Disturbance type	Extra heart beats	Duration of heart rate increase	Mean increase (beats/ min)	*n*	Extra heart beats	Duration of heart rate increase	Mean increase (beats/ min)	*n*
Persons	110	363	18	*16*	173	454	23	*15*
Horse riders	89	387	14	*12*	137	607	14	*8*
Hang-glider	40	183	13	*6*	45	253	11	*6*
Feeding	525	1118	28	*9*	174	579	18	*8*
Gunshot	47	214	13	*10*	14	64	13	*7*
Other acoust.	32	245	8	*12*	29	268	6	*14*

Heart rate reactions were higher at the beginning of the test period (first nine days) than at the end (last nine days), both for optical and acoustical disturbances (Table 32). The decline in extra heart beats result mainly from less magnitude of heart rate increase, only to a lesser extent from a shorter duration of the disturbance effect. However only the trial type "horse riding" had a significant ($p<0.001$) decrease in heart rate reactions with advancing number of disturbance trials (regression equation … $y = 793 - 3.5x$; coefficient of correlation … $r^2 = 0.304$).

Table 32. Heart rate reactions after disturbances of red deer for time period 1 (Jan.7th to Jan. 16th) and time period 2 (Feb. 27th to Mar. 8th)

	Period 1				Period 2			
Disturbance type	Extra heart beats	Duration of heart rate increase	Mean increase (beats/ min)	*n*	Extra heart beats	Duration of heart rate increase	Mean increase (beats/ min)	*n*
Optical	135	395	21	*23*	42	339	8	*18*
Acoustical	36	195	11	*19*	16	158	6	*11*

If disturbance trials with persons walking the round tour are evaluated separately for its subgroups the heart rate reactions are much higher when a group of persons (at least 3 people) walks the tour than a single person (only marginally significant, $p<0.05$). There was no difference between a single person with a dog on leash and a person without dog (Table 33). If a rider gallops past the deer, heart rate increase is higher than with a single rider or group of riders staying at a low pace; these differences were statistically not significant.

Table 33. Heart rate reactions of red deer (2 individuals) after disturbances by persons walking separated for three subgroups and for horse riding trials separated for three subgroups

Disturbance type	Number of additional heart beats	Duration of increased heart rate (in seconds)	Magnitude of heart rate increase (beats/min)	*n*
Single person	103	387	16	*16*
Person with dog	117	344	20	*12*
Group of persons	440	766	34	*3*
Single horse walking	62	413	9	*10*
Single horse trot/gallop	168	466	22	*7*
Group of horse riders	122	703	10	*3*

4. Discussion

Aim of this study was to measure and analyze the effects of anthropogenic disturbances on the behavior and heart rate of roe and red deer under controlled conditions. An essential precondition for this purpose is the knowledge of the natural circadian and seasonal activity rhythms and the corresponding heart rates of the undisturbed animals.

4.1. Circadian and Seasonal Rhythms of Activity and Heart Rate

Roe Deer

Roe deer had a conspicuous period of inactivity at daytime in late summer and early autumn. They seem to have a strong need for resting during daytime in this time of the year and spent most of the time lying in hiding cover. In late autumn the roe deer were much more active during daylight time.

This change to increased daytime activity was quite abrupt. Both roe deer used in the disturbance trials changed their activity pattern on October 21st. Several factors were taken into account as cause. The only temporally concordant incident was a sudden temperature drop on October 21st, in which the primary temperature range of 9°C to 19°C declined to a range of -3°C to 8°C. Presumably this change in activity triggered by the weather change is based on an endogenic process, since all other roe deer in this investigation showed similar activity patterns, although the exact date of activity change differed each year, depending on ambient temperature.

Also heart rate fluctuated strongly between daytime and nighttime in early autumn. The distinct variations in early autumn may indicate a high sensitivity of the cardiovascular system. This can partly be related to shedding hair, and shifting towards winter metabolism.

During rut and post rut roe deer, particularly bucks showed a negative correlation between ambient temperature and the ratio of daytime to nighttime activity. In late summer and autumn roe deer shifted their activity into the night on warm days. In springtime marking and defending the territories and later on fawning seems to overrule these temperature effects. In winter on the other hand, roe deer make use of most of the short daylight time for activity, regardless what temperature. This also has consequences for hunting in summer and late autumn: the chances of observing roe deer during daytime is increased on cooler days.

Male roe deer tended to be less active than females. Mean activity was 46% for females and 40% for male roe deer. Seasonal fluctuations of mean daily activity were very dependent on the individual. One female roe deer in the large enclosure, which approximates the normal home-range size of roe deer, showed strong seasonal fluctuations with highest activity around May (63%) and lowest activity in winter (41%), one buck in the large enclosure varied between 35% in September/October and 49% in April. Guthörl (1994) measured about 49% activity in three free-ranging female roe deer, without any pronounced seasonal variations. Wallach et al. (2010) also could not observe significant differences between activity of roe deer in a Mediterranean environment in summer (47%) and winter (48%). Cederlund (1989) in contrast observed strong seasonal variations in activity of northern boreal roe deer ranging from 30-40% in winter to 57% in late May-early June. Chapman et al. (1993) found quite high mean annual activity rates of 56% for roe deer.

Red Deer

Red deer were more active in summer and autumn than in winter. In general red deer was slightly more active at daytime than at night, especially in winter and spring. Contrary to most investigations including this one, Kamler et al. (2007) could not observe increased activity of red deer during twilight. They explain this with the absence of human hunting, negligible human activities and the presence of large carnivores in the study area.

Heart rate was highest during calving and lactation time and was markedly lower in winter. Not only heart rate but also peripheral body temperature was reduced in cold winter periods (Arnold et al., 2003). Seasonal changes in heart rate are often described in literature. Freddy (1984) measured lowest heart rate in mule deer end of January, and Moen and Chevalier (1977) observed lowest heart rates in white-tailed deer in February. Lieb and Les Marcum (1979) found a negative correlation of heart rate and ambient temperature in wapitis.

4.2. Lunar Cycles

Findings on lunar rhythms in deer are rather inconsistent. Reasons may be different methods for measuring activity (observability vs. biotelemetry), choice of time spans (morning hours, daylight hours, days near full moon or longer phases) or different habitats. Some authors mention increased night activity of roe deer near full moon (Kurt, 1991, Prior, 1968, cited in Turner, 1980). Other studies could not confirm a noticeable effect of the lunar cycle on roe deer activity (Turner, 1980; Pagon, 2010).Whereas Kjær et al. (2008) noticed increased contact rates near full moon, Beier and McCullough (1990) and Webb et al. (2010) could not find any influence of moonlight on nighttime activity or habitat use of white-tailed deer. Buss and Harbert (1950) found a strong correlation between lunar phases and occurrence of mule deer at a salt lick, but both Hayes and Krausman (1993) and Kufeld et al. (1988) observed no differences in activity of mule deer between full and new moon phases. According to Bonchev (2010) the moon affects the reproduction of red deer.

For roe deer no significant differences between activity at full moon and new moon could be found in this study. But red deer showed significant increase of activity at night during full moon and highly significant increase of activity in the morning during new moon phases.

4.3. Disturbance Trials

Roe deer and red deer had quite different reactions towards anthropogenic disturbances.

4.3.1. Impact of Stimuli Types

Roe Deer

Groups of persons created the strongest heart rate and behavioral reactions in roe deer, followed by persons with dogs and gunshots. Reactions towards single persons and horse riders were less and of all optical stimuli tractors elicited the least reactions. Acoustical stimuli other than gunshots hardly provoked any reactions. Roe deer were easily activated by hang-gliders and feeding, but heart rate reactions were only mild. Regarding the single disturbance stimuli types, different behavioral categories were affected in various ways. Locomotion was more pronounced after persons walked by, standing alert after hang-gliders.

Red Deer

Similar to roe deer, persons caused greatest activation and heart rate reactions in red deer. Only feeding caused higher heart rate reactions in the female calf, which will be discussed later on. Horse riders had less impact on behavior and heart rate than persons on foot. Hang-gliders and tractors provoked only mild reactions, and acoustical stimuli other than gunshots had very little influence. Heart rate reactions after gunshots were mild in red deer, which was in contrast to the strong behavioral reactions.

Species Comparison (Parallels, Differences, and General Reaction Patterns of the Species after Disturbances)

Some parallels could be found in the behavior of roe and red deer after disturbances. For both species optical stimuli had a by far higher influence on heart rate and behavior than acoustical stimuli. This result is in accordance with disturbance trials that Hoffmeister (1979) conducted with roe and fallow deer. An exception is the gunshot which elicited very strong heart rate reactions, even though the incidence lasted only a split second, compared to one to three minutes for the other acoustical stimuli. The studied animals were non-hunted, but at single days of each year the surplus of the population was reduced through gun shots. Herbold (1990) could also observe higher flight distances of roe deer after gunshots than through disturbances caused by forest workers or recreationists. The mild heart rate reactions in red deer after gunshots could be a sign of bradycardia, since in more than half of all cases red deer were activated by gunshots, and their behavior was still significantly impacted after 10 minutes (less food intake and more standing alert). Ward and Cupal (1979) in contrast measured a dramatic increase of heart rate in wapiti after gunshots.

After optical disturbances roe deer mostly reacted with standing alert, red deer with increased locomotion. The reactions of roe deer were intense but short, the reactions of red deer longer-lasting. Roe deer showed normal behavior within seven minutes after optical disturbances. After acoustical stimuli they showed the same share of time browsing as undisturbed animals within five minutes, but still spent more time standing alert. Red deer still showed disturbed behavior, both for optical as well as acoustical stimuli, 10 minutes after the disturbance trial. Activated red deer did not compensate the shortened resting bouts, not even in the following night, whereas roe deer rested more after disturbances. Sibbald et al. (2011) observed that the disturbance effect on red deer caused during days with many recreationists tended to last over full 24 h, even when the following night was quiet. In contrast farmed red deer investigated by Hodgetts et al. (1998) returned to pre-stimulus levels of alertness 15 minutes after presenting a disturbance stimulus, and park deer observed by Langbein and Putman (1992) only showed very short-term reactions when disturbed (but it is not mentioned if the population is hunted).

For roe deer both behavioral and heart rate reactions were higher when a person was accompanied by a dog. For red deer there were no differences between a single person walking the tour with a dog on leash or walking without a dog. It should be mentioned that in the present study the dog used in the disturbance trials was a dachshund, and a larger dog breed might have a different effect on the animals. According to Ueckermann et al. (1975) dogs accompanying persons do not effect red deer, fallow deer or wild boar considerably, but in most cases agitate roe deer. This may be that the smaller roe are more prone to fall prey to feral canines. In other studies dogs accompanying a person did not present a special threat in opposition to loose canines, which can cause maximal heart rates and intense flight reactions (for bighorn sheep MacArthur et al., 1979; for roe deer Stadler, 1985; Gaisbauer, 1988). In contrast to these findings, Miller et al. (2001) measured greater flush distances of mule deer when confronted with pedestrians accompanied by a dog on leash than with persons walking alone. And Scottish red deer were more vigilant when dogs were present (Jayakody et al., 2008). Yet many studies indicate that humans evoked flushing of wild ungulates more easily than canines (Stankowich, 2008). After persons walked the round tour with and without a dog, red deer spent much time nosing the ground (olfactory control). In doing so they often walked along the route the persons went. Interestingly the red deer hardly showed this behavior after horses were ridden along the route. It might be that the hooves of the horses leave less traces of scent behind than the soles of dogs or shoes. As mentioned in the methods section, olfactory control was investigated only for red deer (since this behavior was not obvious with roe deer).

Both for roe and red deer reactions towards horse riders were less than towards persons walking without horse. At the first encounter with horse riders the roe deer reacted with intense flight, whereas red deer were much less disturbed compared to persons on foot, even though they never were confronted with horses before. It was irrelevant for the reactions if the horse was of white or brown color. Ueckermann et al. (1975) state that single riders riding quietly do not disturb deer, but agitate them when moving in large groups and fast paces, even when using the familiar trails. These trends could also be observed in this study, but could not be statistically verified (small sample). Kucera (1976) observed shorter flushing distances of white-tailed deer towards horse riders compared with persons on foot. Depending on circumstances roe deer can show very different reactions towards horse riders. In regions with frequently used riding trails roe deer often only stand alert when encountering horse riders without fleeing and it is possible to pass the animals much nearer than on foot, sometimes by only a few meters. In areas, where horse riders are a rare occasion and other disturbance factors such as free-roaming dogs are present, roe deer may panic when confronted with horses (personal observations).

The reactions towards tractors were minimal. This is in contrast to Mrlik (1987), who observed high flight distances of roe deer when confronted with tractors and other motor vehicles. Yet most studies show that wild ungulates reacted less toward vehicles than toward humans on foot (Schultz and Bailey, 1978; MacArthur, 1979; Ward and Cupal, 1979; Kucera, 1979; Stankovich, 2008). Only if a vehicle stopped nearby the animals, they were alerted (Ward and Cupal, 1979). Also all-terrain vehicles traveling off-roads caused stronger reactions (Naylor, 2009).

In contrast to the roe deer, red deer were barely disturbed by the hang-glider. Once by chance a hot air balloon crossed the enclosure very low and the red deer reacted merely with alertly watching it while lying. Also Delap (cit. Heidemann, 1973) mentions the total indifference of red deer towards objects above them. The reason seems to be, that birds of prey do not present a threat for red deer. Roe deer showed increased heart rate reactions after disturbance trials with the hang-glider, even though only the fawns are vulnerable for birds of prey (Raesfeld et al., 1985). Chamois, however, which can be prey to eagles, are very disturbed by hang-gliders, sailplanes and mock eagles (Hamr, 1988).

The high heart rate reactions of the red deer calf during feeding were surprising. This was probably caused by restlessness at the feeding site when the beets thrown in lay near to each other. The male deer chased the calf away which trotted to and fro before it dared to approach the beets again. In practice it is important when feeding red deer to have sufficient feeding sites with ample space between them to avoid social conflicts. Animals low in social rank that must wait at too small feeding stations until the other animals clear the site are also more inclined to bark peeling (Onderscheka, 1986).

In one case the male red deer spike at first fled from the group of people during a disturbance trial but then lay down in a corner of the enclosure and laid its head flat on the ground. As the group drew nearer it got up and continued its flight. Müller-Using and Schloeth (1967) describe adult cervids that are not integrated in a herd going in a prone position when alarmed. Walther (1969) observed two cases where adult Thomson's gazelles chased by dogs abruptly lay down, but he adds that this behavior is an exception. Also Kurt (1991) and Danilkin (1996) mention that adult roe deer can show prone responses towards very strong disturbance stimuli. According to Stöhr (1982) bradycardia combined with passive behavior is an indication for indecisiveness towards an overpowering stimulus.

4.3.2. Influence of Season and Time of Day

Both roe and red deer were more easily activated and stayed longer active in the morning and evening hours. Roe deer showed a strong endogenic activity pattern in early autumn that was barely influenced by disturbance trials, even when frequently carried out. In this time period the roe deer had an intense want for resting at daytime and could hardly be activated through optical or acoustical stimuli. If they had to escape, they were active for only a short time span and the animals mostly lay down again within a couple of minutes. Such short interruptions could split a resting bout, increasing the number of resting bouts per day and shortening the mean duration of the bouts, yet the total activity budget remained uninfluenced by the disturbances. Analysis of their behavior showed that the investigated roe deer mostly stood alert or showed locomotion during the short active phase, thus they could be considered as disturbed until they lay down again.

In late autumn roe deer reacted stronger towards disturbances, yet this should not necessarily be interpreted as sensitization towards disturbances, since the roe deer generally tended to be much more active in this time of season, also when not disturbed. If a lying deer was activated in late autumn it stayed longer active before lying down again, sometimes even for hours. However they did not show disturbed behavior during the whole active time period. The disturbance seemed to be more a stimulus for changing their activity, as Heidemann (1973) already stated.

Red deer were investigated in disturbance trials only in winter, so that no explicit statements about seasonal variation in disturbance effects can be made.

4.3.3. Compensation Strategies

Roe and red deer showed different strategies of compensating after being disturbed. Both in early and late autumn roe deer did not change the amount of total activity in the daytime as well as for 24 hours when disturbed. The length of resting bouts of roe deer were shortened on days with disturbance trials. At the same time the number of resting bouts per day increased, so that the total activity budgets for daytime as well as for 24 hours were not influenced by disturbance trials.

The red deer, when activated, stayed in average nearly 40 minutes active before lying down again. As a consequence of the activations red deer shortened their resting bouts, but instead of compensating the loss of resting time by increasing the number of resting bouts such as roe deer did, they increased the daily amount of time active. Thus their activity budget was modified by disturbances, on days with disturbance trials they rested half an hour less in the daytime and did not compensate this in the night.

In spite of the very short and strong heart rate reactions after disturbances, roe deer did not show any long-term changes in resting heart rate. Red deer in contrast had decreased resting heart rates on days on which disturbance trials were carried out. Possibly the animals compensate the increased proportion of activity on days in which they are often disturbed with subsequent deeper resting phases. Another explanation would be in raised attentiveness on restless days which is accompanied by bradycardia.

4.3.4. Habituation

Towards the end of the test period red deer showed a decreased duration of disturbed behavior and less heart rate reactions, which suggest a certain habituation towards optical and acoustical stimuli (but not significant). Yet solely with horse riding trials a significant reduction of heart rate reaction with proceeding disturbance trials could be observed. The responsiveness of roe deer towards disturbances seems to be strongly dependent on the season. In the second half of the test period (late autumn) roe deer were much more susceptible to disturbances as in the first half of the test period (early autumn). Therefore no habituation could be determined for the investigated time period.

4.3.5. Synchronization of Behavior

Between both red deer stronger synchronization of activity in the disturbances trials could be found than between the roe deer. One explanation could be the strong social structure of red deer that may lead to more uniform reactions. However, it must be pointed out that both deer species were investigated in disturbance trials in different seasons, therefore a direct comparison is not possible. Several investigations showed stronger synchronization of behavior during the winter months for roe deer (Ellenberg, 1978; Mertens, 1984; Jeppesen, 1989) and red deer (Georgii and Schröder, 1981). Jeppesen (1989) attributes this to the stronger photoperiodic influence of behavior in winter. Thus synchronization of behavior is also much stronger during twilight than at daylight time (Turner, 1980). According to Bützler (1974) specific behavioral types, such as flight or standing alert, are stronger synchronized than others (e.g. comfort behavior).

4.3.6. Bedding Sites

Roe Deer

The sudden change of bedding site of the roe deer on October 24th was probably caused by spanning the rope for the hang-glider across the small grass patch where they usually had lain. The weather change and the change in activity pattern already took place beforehand on October 21st, and the third roe deer was introduced to the enclosure later on. The roe deer mostly chose bedding sites in which they were partly sheltered by high grasses or shrubs. According to Mysterud and Østbye (1995) winter bedding sites of roe deer in open habitat required ground cover for protection against the wind and no beds exposed to direct sunlight were recorded in the "summer" study of Mysterud (1996).

Red Deer

It was noticeable that the red deer investigated often chose the same bedding site again on one day, even when disturbed, but selected a different bedding site on the next day. Possibly weather conditions influenced their choice. According to Heidemann (1973) shelter from wind was crucial for fallow deer in selecting the bedding site and the red deer investigated by Fischer and Gossow (1985) preferred bedding sites under cover of trees. These tendencies could not be perceived with the red deer investigated in the trials, after all in almost half of all bedding cases they lay on the open grass field, partly even on the small, sun exposed mound with good overview. This was in accordance with the red deer Zhang and Xiao (1990) studied, which selected bedding areas exposed to the sun during late winter. Jiang et al. (2007) frequently found red deer bedded in shrub stands near sunny ridge tops and stated that human disturbances had only little discernible effect on bedding site selection. Also Millspaugh et al. (1998) found that wapitis were not influenced by predictable human disturbances in choosing their bed sites, but rather sought places with adequate thermal cover in summer.

4.3.7. Methodological Aspects

Activation and length of active bouts after being activated could easily be identified from the offline-data (based on minutes). Activation appears to be a better indicator for the influence of anthropogenic disturbances than the duration. With roe deer it was insignificant for the length of the following active bout by which type of disturbance it was induced.

Rather than comparing the active behavior after disturbance with the resting behavior before, it made sense to compare the behavior of activated animals with undisturbed active behavior and measure how long it takes until normal active behavior occurs again. In doing so food intake and standing alert should particularly be taken in account, since food intake is important for the energy budget of the animal and standing alert is an indication of disturbance. The share of time standing alert is a better criterion for disturbance than the frequency of lifting the head. According to these criteria roe deer showed normal behavior within seven minutes after optical disturbances.

Heart rate is a good indicator for disturbance effects in animals and owing to its close coherence to energy budget and the relatively easy measurability it is a popular parameter for examining stress in animals. Thereby the number of additional heart beats proved to be a better measure for a disturbance effect than the maximal or mean increase in heart rate. Even so, heart rate, if used by itself, is not always conclusive, since it is highly dependent on the respective activity of the animal. In some situations heart rate actually drops when the animal is disturbed, e.g. during standing alert (alarm bradycardia). Therefore it makes sense to combine measurements of heart rate with behavioral analysis.

The results of the disturbance trials show the effects of various disturbances on heart rate, activity and behavior of roe and red deer under controlled, experimental conditions. However, the results cannot be simply transferred to free-ranging animals. Following points should be considered:

- The animals were raised in captivity and were to some extent accustomed to humans, although they were not tame. Entirely new stimuli for the animals were horses and hang-gliders.
- The enclosure limited the possibility for the animals to escape from the source of disturbance, so that the reaction provoked by the stimulus was possibly prolonged.
- The hang-glider always “flew” the same stretch. After a few trials the surprise effect of an object suddenly appearing in the sky from any direction is reduced (spatially specific habituation).
- The population was non-hunted (except single days per year for regulation culling). Experimental gunshots did not have a negative outcome for the animals. Bützler (2001) indicates that red deer integrated in a herd link the sound of gunshots with a culled animal and maybe even pass on this experience to the next generation. Hunted roe deer also had greater flight distances than non-hunted animals (de Boer, 2004).
- It should also be mentioned that the tractor was driven outside of the enclosure, whereas the disturbance trials with persons or horses took place inside the fenced area. The separating fence may have an influence on the flight reactions of the animals.
- The animals were fed ad libitum. This has an influence on the physical condition of the animals and the resultant behavior, for instance reindeer in bad physical condition reacted differently towards hunters than well-fed animals (Skogland and Grøvan, 1988). Undernourished and well-fed white-tailed deer had similar seasonal and circadian activity patterns, yet the undernourished animals showed much stronger flight reactions (Ozoga and Verme, 1970).
- Only two young red deer were tested in the trials. According to Petrak (1988) red deer groups including older and experienced animals often

reacted only with standing alert when disturbed, whereas groups with young, inexperienced animals fled in the same situation.

Disturbance trials in enclosures also have many advantages. The good observability facilitates the interpretation of the biotelemetrically transmitted data, since it could repeatedly be compared with video recordings. Above all in the wild there are frequently uncontrolled disturbances, which hinder the interpretation of activity and heart rate.

CONCLUSION

Disturbances by humans can cause changes in behavior, activity pattern and heart rate of deer. Optical stimuli generally caused high short-time heart rate reactions, whereas acoustic stimuli evoked only mild heart rate reactions, except for gunshots, which elicited strong heart rate reactions both in red deer and especially in roe deer. Altogether the reactions of roe deer towards stimuli were short and strong, whereas red deer tended to show less obvious but persisting reactions.

Roe deer were much more susceptible to disturbances in times when they showed more activity (twilight, late autumn). It can be assumed, that in the rest of the year the animals are also more easily disturbed by human activities at times when they themselves are more active. Disturbances in winter are in double sense a concern. First, deer react more sensitive towards disturbance stimuli in this time of the year, and secondly the impact on their energy budget is of greater consequence in winter with food shortage. This should be specifically considered with human winter recreation activities, particularly at dawn and dusk.

Hunting also has a great influence on the reaction of deer towards disturbances. Non-hunted populations have less flight responses than hunted populations (Schultz and Bailey, 1978; de Boer et al., 2004; Stankowich, 2008). Deer shift their home range during hunting season and increase nocturnal activity (Kilgo et al., 1998). With progressing hunting season deer changed behavior and visibility of deer decreased making hunting effort more difficult (Grau and Grau, 1980). Thus hunting seasons should be kept short, if disturbances of deer are to be minimized. This can be realized by interval hunting (Reimoser, 1991). Focus hunting (Reimoser, 1991) on the other hand can be targeted to keep deer out of areas susceptible to game damage, by culling higher numbers and deliberately creating disturbance to drive the animals in other less sensitive areas.

Especially red deer are less disturbed by horse riders than by persons walking alone or in groups. This may be that horses as herbivores are of no threat to deer and partly cover the scent of the rider or at least signalize that there is no threat of other predators nearby. This would not be the case if the deer were hunted from horseback. Also wild animals tend to respond most to the human form (Taylor and Knight, 2003), which is concealed on horseback. Wildlife biologists can even take advantage of this fact and benefit from a better viewpoint, and easier transport of equipment in areas inaccessible by vehicles (Kucera, 1976). And disturbance by horseback riding is minimal compared to all-terrain vehicles (Naylor et al., 2009).

Dogs should stay directly nearby to persons (e.g. on leash), then they do not affect deer considerably more than persons without dogs, contrary to free-running dogs. Walking off trails leads to more disturbance of deer than staying on paths (Herbold, 1995; Miller et al., 2001; Taylor and Knight, 2003) and deer flee much more often when an observer stops, as opposed to walking continuously onward (Borkowski, 2001). People are not always aware of how they affect wildlife and often underestimate their impact (Taylor and Knight, 2003; Sterl et al., 2008). This should be taken into account, as animals may lose large parts of their habitat due to frequent recreational activities (Gander and Ingold, 1997).

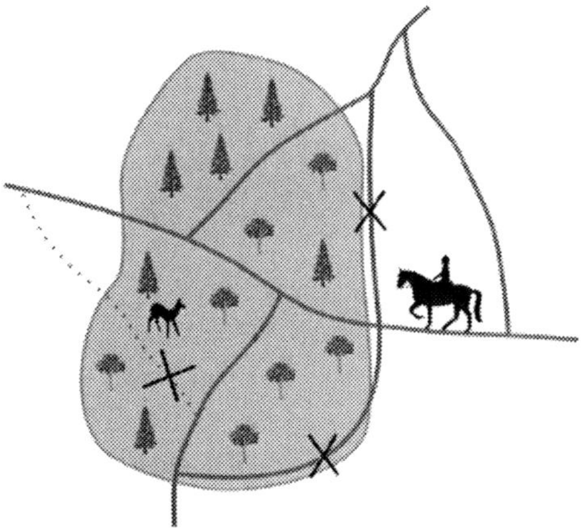

Figure 62. Example of recommendations for recreationists to minimize disturbances. Paths should be used instead of going off-trail. Very attractive areas such as forest-meadow boundaries ("edge effect") or breeding places should be avoided especially in critical time periods.

To which extent is also dependent on the species, e.g. roe deer are behaviorally very flexible (Jepsen and Topping, 2004) and seem to cope with human disturbance near settlements better than red deer (Jiang et al., 2008). In critical habitat areas and wildlife sanctuaries recommendations for recreationists should be given to minimize disturbances (Figure 62), for example using paths instead of going off-trail, avoiding very attractive areas for deer such as forest-meadow boundaries ("edge effect", Reimoser and Ellenberg, 1999) or breeding places (Knight and Cole, 1995a), especially at times when animals are more susceptible to disturbances, e.g. in winter, at dawn or dusk.

In management plans it is also important to distinguish between local, short-term disturbances and regional, long-term disturbances, such as road constructions, and also take cumulative effects into account (Knight and Cole, 1995b; Vistnes and Nellemann, 2008).

ACKNOWLEDGMENTS

I am heartily thankful for the support and mentoring received from Dr. Franz Schober and Dr. Franz Suchentrunk from the Research Institute of Wildlife Ecology, University of Veterinary Medicine Vienna, and the collaboration with Dr. Helmut Herbold in carrying out the disturbance trials. I am also very grateful for the help of Dr. Horst Wagner in programming the actograms.

REFERENCES

Adrados, C., Verheyden-Tixier, H., Cargnelutti, B., Pépin, D., Janeau, G. 2003. GPS approach to study fine-scale site use by wild red deer during active and inactive behaviors. *Wildl. Soc. Bull.* 31 (2): 544-552.

Albrecht, L. 1988. Die Gewöhnung von Gemsen an den Menschen [Habituation of chamois towards humans]. *Wildtiere* 3: 1-4.

Altmann, M. 1958. The flight distance in free-ranging big game. J. *Wildl. Manage.* 22: 207-209.

Apollonio, M., Ciuti, S., Luccarini, S. 2005. Long-term influence of human presence on spatial sexual segregation in fallow deer (Dama dama). *J. Mammal.* 86(5): 937-946.

Archer, S., Stokes, C. 2000. Stress, disturbance and changes in rangeland ecosystems. *In: Rangeland desertification*. Arnalds, O., Archer, S. (Eds.): Dordrecht: Kluwer Academic Publishers. pp. 17-38.

Arnold, W., Ruf, T., Reimoser, S., Tataruch, F., Onderscheka, K., Schober, F. 2003. Nocturnal hypometabolism as an overwintering strategy of red deer (*Cervus elaphus*). *Am. J. Physiol.* 286: R174–R181.

Aschoff, J. 1966. Circadian activity pattern with two peaks. *Ecology* 47: 657-662.

Ball, N.J., Amlaner, Jr. 1979. Changing heart rates of herring gulls when approached by humans. *In: A handbook on biotelemetry and radio-tracking*. Amlaner, C.J., MacDonald, D.W. (Eds.). Oxford: Pergamon Press. pp. 589-594.

Barth, W.-E. 1984. Menschliche Störungen in der Natur II. [Human disturbances in nature]. *Wild und Hund* 10: 37-42.

Beier, P., McCullough, D.R. 1990. Factors influencing white-tailed deer activity patterns and habitat use. *Wildl. Monogr.* 109. 51 pp.

Berger, A., Scheibe, K.-M., Brelurut, A., Schober, F., Streich, W.J. 2002. Seasonal variation of diurnal and ultradian rhythms in red deer. *Biol. Rhythm Res.* 33(3): 237-253.

Bergerud, A.T., Jakimchuk, R.D., Carruthers, D.R. 1984. The buffalo of the north: Caribou (Rangifer *tarandus*) and human developments. *Artic.* 37: 7-22.

Blackshaw, J.K. 1986. Objective measures of welfare in farming environments. *Australian Veterinary Journal* 63: 361-364.

Bonchev, L. 2010. Research on the seasonal variability in the reproductions of red deer (Cervus elaphus L.) on population level and the moon as a geophysical factor affecting that process in Bulgaria. [Abstract in English]. 20th Anniversary International Scientific Conference 3rd - 4th June 2010, Stara Zagora, Bulgaria. Agricultural science. *Animal studies and Veterinary medicine*. 30-34.

Borkowski, J. 2001. Flight behaviour and observability in human-disturbed sika deer. *Acta Theriol.* 46 (2): 195–206.

Boyle, S.A., Samson, F.B. 1985. Effects of nonconsumptive recreation on wildlife: *A review. Wildl. Soc. Bull.* 13: 110-116.

Burch, W. 1988. Human ecology and environmental management. *In: Ecosystem management for parks and wilderness*. Agee, J.K., Darryll, R.J. (Eds.). Seattle and London: University of Washington Press. pp. 145-159.

Buss, I.O., Harbert, F.H. 1950. Relation of moon phases tot he occurrence of mule derer at a Washington salt lick. *J. Mammal.* 31 (4): 426-429.

Büttner, K. 1980. Untersuchungen zum Einfluß von Größe und Zusammensetzung des Wintersprunges beim Rehwild auf die Zeitdauer verschiedener Verhaltensweisen [Investigation of the influence of size and composition of winter herds of roe deer on the duration of various behavioral characteristics]. *Z. Jagdwiss.* 26: 181-194.

Bützler, W. 1974. *Kampf- und Paarungsverhalten, soziale Rangordnung und Aktivitätsperiodik beim Rothirsch* [Fighting and breeding behavior, social hierarchy and activity periodicity of Red Deer]. Berlin: Parey. 80 pp.

Bützler, W. 2001. *Rotwild: Biologie-Verhalten-Umwelt-Hege* [Red Deer: Biology-Behavior-Environment-Gamekeeping]. BLV Jagdbuch, 5th edition. Munich: BLV-Verlagsgesellschaft. 264 pp.

Cannon, W.B. 1929. Bodily changes in pain, hunger, fear and rage: *An account of recent researches into the function of emotional excitement.* Oxford: Appleton. 404 pp.

Cederlund, G. 1981. Daily and seasonal activity pattern of roe deer in a boreal habitat. *Viltrevy* 11: 315-353.

Cederlund, G. 1989. Activity Patterns in Moose and Roe Deer in a North Boreal Forest. *Holarctic Ecology* 12(1): 39-45.

Cederna, A., Lovari, S. 1985. The impact of tourism on chamois feeding activities in an area of the Abruzzo National Park, Italy. *In: The biology and management of mountain ungulates.* Lovari, S. (Ed.). London: Croom Helm. pp. 216-225.

Chapman, N.G., Claydon, K., Claydon, M., Forde, P.G., Harris, S. 1993. Sympatric populations of muntjac (*Muntiacus reevesi*) and roe deer (*Capreolus capreolus*): a comparative analysis of their ranging behaviour, social organization and activity. *J. Zool.* 229 (4): 623–640.

Christensen, N.L. 1988. Succession and natural disturbance: Paradigms, problems, and preservation of natural ecosystems. *In: Ecosystem management for parks and wilderness.* Agee, J.K., Darryll, R.J. (Eds.). Seattle and London: University of Washington Press. pp. 62-86.

Collins, W.B., Urness, P.J., Austin, D.D. (1978): Elk diets and activities on different lodgepole pine habitat segments. *J. Wildl. Manage.* 42, 799-810.

Czech, B. 1991. Elk behavior in response to human disturbance at the Mount St. Helens National Volcanic Monument. *Appl. Anim. Behav. Sci.* 29: 269-277.

Danilkin, A. 1996. *Behavioural ecology of Siberian and European roe deer.* London: Chapman and Hall. 298 pp.

de Boer, H.Y., van Breukelen, L., Hootsmans, M.J.M., van Wieren, S.E. 2004. Flight distance in roe deer *Capreolus capreolus* and fallow deer *Dama dama* as related to hunting and other factors. *Wildl. Biol.* 10(1): 35-41.

Dimond, S., Lazarus, J. 1974. The problem of vigilance in animal life. *Brain Behav. Evolut.* 9: 60-79.

Duscher, A.; Filli, F.; Reimoser, F.; Lainer, F. 2009. *Impact of supplementary feeding on winter home range size and activity patterns of female red deer (Cervus elaphus) in alpine regions.* 4th Symposium of the Hohe Tauern National Park for Research in Protected Areas; Sept 17-19, 2009; Mittersill, Austria. pp. 73-76. (ISBN: 978-3-9502062-1-0).

Ellenberg, H. 1978. Zur Populationsökologie des Rehes (Capreolus capreolus L., Cervidae) in Mitteleuropa [On the population ecology of roe deer in Central Europe]. *Spixiana.* Supplement 2. 211pp.

Espmark, Y., Langvatn, R. 1979. Cardiac responses in alarmed red deer calves. *Behavioral Processes* 4: 179-186.

Espmark, Y., Langvatn, R. 1985. Developement and habituation of cardiac and behavioral responses in young red deer calves (Cervus elaphus) exposed to alarm stimuli. J. *Mamm.* 66: 702-711.

Fancy, S., White, R. 1985. Energy expenditures by caribou while cratering in snow. *J. Wildl. Manage.* 49(4): 987-993.

Ferns, P.N., MacAlpine-Leny, I.H., Goss-Custard, J.D. 1979. Telemetry of heart rate as a possible method of estimating energy expenditure in the redshank. *In: A handbook on biotelemetry and radio tracking.* Amlaner und MacDonald (Eds.). Oxford: Pergamon Press. pp. 595-602.

Fischer, A., Gossow, H. 1985. Zur winterlichen Einstandswahl des Rotwildes in Abhängigkeit von Sicherheits- und Klimaschutzansprüchen. *XVIIth Congress of the International Union of Game Biologists.* Brussels, Sept. 17-21, 1985. pp. 215-222.

Fraser, D., Richie, J.S.D., Fraser, A.F. 1975. The term "stress"' in a veterinary context. *Br. Vet. J.* 131: 653-662.

Freddy, D.J. 1979. Measuring heart rates of mule deer using a repeater-type telemetry system. *Proc. 2nd Int. Conf. Wildl. Biotelem.* pp. 144-155.

Freddy, D.J. 1984. Heart rate for activities of mule deer at pasture. *J. Wildl. Manage.* 48: 962-969.

Freddy, D.J., Bronaugh, W.M., Fowler, M.C. 1986. Responses of mule deer to disturbance by persons afoot and snowmobiles. *Wildl. Soc. Bull.* 14: 63-68.

Frid, A., Dill, L.M. 2002. Human-caused disturbance stimuli as a form of predation risk. Conservation Ecology 6(1): 11. (online) URL: http://www.consecol.org/vol6/iss1/art11/.

Gabrielsen, G.,W., Smith, E.,N. 1995. Physiological responses of wildlife to disturbances. In: *Wildlife and recreationists.* Knight, R.L., Gutzwiller, K.J. (Eds.): Washington D.C.: Island Press. pp. 95-108.

Gaisbauer, R. 1988. Telemetrisch ermittelte Reaktionen des Rehwildes auf simulierte Störungen im Nationalpark Bayerischer Wald [Reactions of roe deer towards simulated disturbances investigated via telemetry in the National Park Bayerischer Wald]. Fachhochschule Weihenstephan. *Diploma thesis.* 113 pp.

Gander, H., Ingold, P. 1997. Reactions of male alpine chamois *Rupicapra r. rupicapra* to hikers, joggers and mountainbikers. *Biological Conservation* 79: 107-109.

Geist, V. 1971. Is big game harassment harmful? *Oilweek* 22: 12-13.

Georgii, B. 1976. Untersuchungen über das Nebennierengewicht des Rehes [Research of adrenal weights of roe deer]. *Z. Jagdwiss.* 22: 85-97.

Georgii, B. 1980. Einflüsse menschlicher Störungen auf Standortwahl und Aktivitätsmuster weiblicher Rothirsche [Influence of human disturbances on habitat selection and activity patterns of female Red Deer]. *Verhandlungen der Gesellschaft für Ökologie*, Freising-Weihenstephan. Band VIII. pp. 163-168.

Georgii, B. 1981. Activity patterns of female red deer (*Cervus elaphus L.*) in the Alps. *Oecologia* 49: 127-136.

Georgii, B. 1984. Activity patterns of free-ranging red deer *Cervus elaphus L. Acta Zool. Fennica* 172: 81-83.

Georgii, B., Schröder, W. 1978. Radiotelemetrisch gemessene Aktivität weiblichen Rotwildes [A radiotelemetric study of the activity of female red deer]. *Z. Jagdwiss.* 24: 9-23.

Georgii, B., Schröder, W. 1981. Raumnutzungs- und Aktivitätsmuster von Rothirschen unter besonderer Berücksichtigung menschlicher Störungen [Range use and activity patterns of red deer especially with consideration of human disturbances]. München: *Wildbiol. Gesellschaft.* 56 pp.

Giacometti, M., Janovsky, M., Fluch, G., Arnold, W., Schober, F. 2001. A Technique to Implant Heart-Rate Transmitters in Red Deer. *Wildl. Soc. Bull.* 29 (2): 586-593.

Gill, J.A., Sutherland, W.J., Watkinson, A.R. 1996. A method to quantify the effects of human disturbance for animal populations. *J. Appl. Ecol.* 33: 786–792.

Gill, J.A., Norris, K., Sutherland, W.J. *2001.* Why behavioural responses may not reflect the population consequences of human disturbance. *Biological Conservation* 97: 265-268.

Grau, G.A., Grau, B.L. 1980. Effects of hunting on hunter effoert and white-tailed deer behavior. *Ohio Acad. Sci.* 80 (4): 150-156.

Green, R.A., Bear, G.D. 1990. Seasonal cycles and daily activity patterns of rocky mountain elk. *J. Wildl. Manage.* 54: 272-279.

Groves, P.M., Thompson, R.F. 1970. Habituation: A dual-process theory. *Psychological Review* 77(5): 419-450.

Guthörl, V. 1987. Radiotelemetrische Untersuchungen zum Raum-Zeit-System des Rehwildes, Capreolus c. capreolus [Radiotelemetric research on the spatial-temporal-system of roe deer]. Institut für Biographie der Universität des Saarlandes. *Diploma thesis.*

Guthörl, V. 1994. Aktivitätsmuster von Rehen (Capreolus capreolus Linné, 1758) in einem Stadtwald mit starkem Erholungsverkehr. [Activity patterns of roe deer in a city forest with high recreational use] *Z. Jagdwiss.* 40: 241-252.

Hamr, J. 1988. Disturbance behaviour of chamois in an alpine tourist area of Austria. *Mountain Research and Developement.* 8: 65-73.

Harlow, H.J., Thorne, E.T., Williams, E.S., Belden, E.L., Gern, W.A. 1987. Cardiac frequency: a potential predictor of blood cortisol levels during acute and chronic stress exposure in Rocky Mountain bighorn sheep (*Ovis Canadensis Canadensis*). *Can. J. Zool.* 65: 2028-2034.

Hayes, C.L., Krausman, P.R. 1993. Nocturnal Activity of Female Desert Mule Deer. *J. Wildl. Manage.* 57 (4): 897-904.

Heidemann, G. 1973. Zur Biologie des Damwildes. (Biology of fallow deer) In: *Mammalia depicta* 9. Hamburg: Parey. 95 pp.

Herbold, H. 1990. Reaktion von Rehwild auf Störungen durch Menschen. [Reaction of roe deer to disturbances by humans]. *Trans. 19th IUGB Congress.* Trondheim 1989. pp. 414-420.

Herbold, H. 1995. Raum-Zeit-Verhalten von Rehen *capreolus capreolus* in anthropogen beeinflussten Einständen [Spatio-temporal activity of Roe Deer in man-influenced ranges]. *Der Ornithologische Beobachter* 92: 275-280.

Hernández, L., Laundré, J.W. 2005. Foraging in the "landscape of fear" and its implications for habitat use and diet quality of elk *Cervus elaphus* and bison *Bison bison. Wildl. Biol.* 11: 215-220.

Hewison, A.J.M., Vincent, J.P., Joachim, J., Angibault, J.M., Cargnelutti, B., Cibien, C. 2001. The effects of woodland fragmentation and human activity on roe deer distribution in agricultural landscapes. *Can. J. Zool.* 79: 679–689.

Hodgetts, B.V., Waas, J.R., Matthews, L.R. 1998. The effects of visual and auditory disturbance on the behaviour of red deer (*Cervus elaphus*) at pasture with and without shelter. *Appl. Anim. Behav.* Sci.55: 337-351.

Hoffmeister, H. 1979. Untersuchungen über die Reaktionen von Reh- und Damwild auf verschiedene Umwelteinflüße unter Einsatz der telemetrischen Erfassung von Herzfrequenz und EKG. [Investigating reactions of roe and fallow deer to various environmental impacts by telemetrically measuring heart rate and ECG]. Hannover. Veterinary University. *Thesis.*

Holter, J.B., Urban, W.E., Hayes, H.H., Silver, H. 1976. Predicting metabolic rate from telemetered heart rate in white-tailed deer. *J. Wildl. Manage.* 40: 626-629.

Hood, R.E., Inglis, J.M. 1974. Behavioral responses of white-tailed deer to intensive ranching operations. *J. Wildl. Manage.* 38: 488-498.

Horejsi, B. 1976. Some thoughts and observations on harassment and bighorn sheep. *Biennial Meeting of the Northern Wild Sheep Council*, Jackson, Wyoming, Febr. 10-20, 1976. pp. 1-14.

Huber, S., Palme, R., Zenker, W., Möstl, E. 2003. Non-invasive monitoring of the adrenocortical response in Red Deer. *J. Wildl. Manage.* 67(2): 258-266.

Ingold, P. (2005): *Freizeitaktivitäten im Lebensraum der Alpentiere* [Outdoor recreation in the habitat of Alpine wildlife]. Bern: Haupt. 516 pp.

Jaeger, J. A. G., Bowman, J., Brennan, J., Fahrig, L., Bert, D., Bouchard, J., Charbonneau, N., Frank, K., Gruber, B., Tluk von Toschanowitz, K. 2005. Predicting when animal populations are at risk from roads: an interactive model of road avoidance behavior. *Ecol. Mod.* 185: 329–348.

Jacobsen, Nadine K. 1979a. Changes in heart rate with growth and activity of white-tailed deer fawns. *Comp. Biochem. Physiol.* 62A: 885-888.

Jacobsen, Nadine K. 1979b. Alarm bradykardie in white-tailed deer fawns (*Odocoileus virginianus*). *J. Mamm.* 60: 343-349.

Jayakody, S., Sibbald, A., Gordon, I.J., Lambin, X. 2008. Red deer *Cervus elephus* vigilance behaviour differs with habitat and type of human disturbance. *Wildl. Biol.* 14: 81-91.

Jayakody, S., Sibbald, A., Mayes, R.W., Hooper, R.J., Gordon, I.J., Lambin, X. 2011. Effects of human disturbance on the diet composition of wild red deer (Cervus elaphus). *Eur. J. Wildl. Res.*, 10 pp. (DOI 10.1007/s10344-011-0508-z)

Jenkins, W.L., Kruger, J.M. 1973. Modern concepts of the animal's physiological response to stress. *In: The capture and care of wild animals.* Young, E. (Ed.): Hollywood: Ralph Curtis books. 172-182.

Jeppesen, J.L. 1984. Human disturbance of roe deer and red deer: Preliminary results. *Commun. Inst. For. Fenn.* 120: 113-118.

Jeppesen, J.L. 1987. Impact of human disturbance on home range, movements and activity of red deer in a Danish environment. *Dan. Rev. Game Biol.* 13: 1-38.

Jeppesen, J.L. 1989. Activity patterns of free-ranging roe deer (*Capreolus capreolus*) at Kalø. *Dan. Rev. Game Biol.* 13: 1-23.

Jepsen, J.U. , Topping, C.J. .2004. Modelling roe deer (*Capreolus capreolus*) in a gradient of forest fragmentation: behavioural plasticity and choice of cover. *Can. J. Zool.* 82 (9): 1528-1541

Jiang, G., Ma, J., Zhang, M., Ma, J. 2007. Effects of human disturbance on movement, foraging and bed selection in red deer *Cervus elaphus xanthopygus* from the Wandshan Mountains, northeastern China. *Acta Theriol.* 52 (4): 435-446.

Jiang, G., Ma, J., Zhang, M., Ma, J. 2008. Habitat use and separation between red deer *Cervus elaphus xanthopygus* and roe deer *Capreolus pygargus bedfordi* in relation to human disturbance in the Wandashan Mountains, northeastern China. *Wildl. Biol.* 14(1): 92-100.

Jiang, G., Ma, J., Zhang, M., Stott, P. 2009. Effects of human activities on the spatial distribution of eastern roe deer *Capreolus pygargus bedfordi* in the Lesser Khingan Mountains, northeastern China. *Acta Theriol.* 54 (1): 61–76.

Jones, A.R., Price, S. 1990. Can stress in deer be measured? *Deer* 8: 25-27.

Kamler, J.F., Jędrzejewska, B., Jędrzejewski, W. 2007. Activity patterns of red deer in Białowieża National Park, Poland. *J. Mammal.* 88(2): 508–514.

Kammermeyer, K.E., Marchinton, R.L. 1977. Seasonal change in circadian activity of radio-monitored deer. *J. Wildl. Manage.* 41: 315-317.

Keay, J. M., Singh, J., Gaunt, M. C., Kaur, T. 2006. Fecal glucocorticoids and their metabolites as indicators of stress in various mammalian species: a literature review. *J. Zoo Wildl. Med.* 37: 234–244.

Kilgo, J.C., Labisky, R.F., Fritzen D.E. 1998. Influences of Hunting on the Behavior of White-Tailed Deer: Implications for Conservation of the Florida Panther. *Conserv. Biol.* 12(6): 1359–1364.

Kjær L.J., Schauber, E.M., Nielsen, C.K. 2008. Spatial and Temporal Analysis of Contact Rates in Female White-Tailed Deer. *J. Wildl. Manage.* 72 (8): 1819-1825.

Knight, R.L., Cole, D.N. 1995a. Factors that influence wildlife responses to recreationists. *In: Wildlife and recreationists*. Knight, R.L., Gutzwiller, K.J. (Eds.). Washington D.C.: Island Press. pp. 71-79.

Knight, R.L., Cole, D.N. 1995b. Wildlife responses to recreationists. *In: Wildlife and recreationists*. Knight, R.L., Gutzwiller, K.J. (Eds.). Washington D.C.: Island Press. pp. 51-69.

Knight, R.L., Gutzwiller, K.J. 1995. *Wildlife and recreationists*. Washington D.C.: Island Press. 389 pp.

Kreeger, T.J. 1990. Physiological monitoring of gray wolves (*Canis lupus*) by radiotelemetry. *J. Mamm.* 71: 258-261.

Kucera, E. 1976. Deer flushing distance as related to observer's mode of travel. *Wildl. Soc. Bull.* 4 (3): 128-129.

Kufeld, R.C., Bowden, D.C., Schrupp, D.L. 1988. Habitat Selection and Activity Patterns of Female Mule Deer in the Front Range, Colorado. *J. Range Manage.* 41 (6): 515-522.

Kurt, F. 1991. *Das Reh in der Kulturlandschaft: Sozialverhalten und Ökologie eines Anpassers* [Roe deer in the cultural landscape: Social behavior and ecology of an adaptor]. Hamburg: Parey. 288pp.

Langbein, J., Putman, R.J. 1992. Behavioural reponses of park red and fallow deer to disturbance and effects on population performance. *Animal welfare* 1: 19-38.

Lark, A.,M., Slade, N.,A. 2008. Variation in Vigilance in White-tailed Deer (*Odocoileus virginianus*) in Northeastern Kansas. *Am. Midl. Nat.* 159: 67–74.

Levine, S. 1985. A definition of stress? *In: Animal stress*. Moberg, G.P. (Ed.): Baltimore: Waverly Press. pp. 51-69.

Lieb, J.W. 1981. Activity of free-ranging elk as determined by a motion-sensing biotelemetry system. *Proc.3rd Int. Conf. on Wildl. Biot.* 1-12.

Lieb, J.W., Les Marcum, C. 1979. Biotelemetric monitoring of heart rate and activity in elk. *Proc. 2nd Int. Conf. on Wildl. Biot.*, 21-32.

MacArthur, R.A., Johnston, R.H., Geist, V. 1979. Factors influencing heart rate in free-ranging bighorn sheep: A physiological approach to the study of wildlife harassment. *Can. J. Zool.* 57: 2010-2021.

MacArthur, R.A., Geist, V., Johnston, R.H. 1982. Cardiac and behavioural responses of mountain sheep to human disturbance. *J. Wildl. Manage.* 46: 351-358.

Manor, R. Saltz. D. 2003. Impact of human nuisance disturbance on vigilance and group size of a social ungulate. *Ecological Applications* 13: 1830–1834.

Mautz, W., Fair, J. 1980. Energy expenditure and heart rate for activities of white-tailed deer. *J. Wildl. Manage.* 44: 333-342.

Mertens, C. 1984. Zur Verhaltensynchronisation bei gruppenweise gehaltenen Rehen (*Capreolus capreolus*) [Behavioral synchronization of roe deer held in groups]. *Behavior* 89: 220-240.

Miller, N.E. 1982. Motivation and psychological stress. *In:* The physiological mechanisms of motivation. Pfaff, D.W. (Ed.). New York: Springer. pp. 409-432.

Miller, F.L., Gunn, A. 1984. Muskox defense formations in response to helicopters in the Canadian high artic. *Biol. Pap. Univ. Alaska Spec. Rep.* 4: 123- 126.

Miller, S.G., Knight, R.L., Miller, C.K. 2001. Wildlife responses to pedestrians and dogs. *Wildl. Soc. Bull.* 29: 124-132.

Millspaugh J. J., Raedeke K. J., Brundige G. C., Willmott C. C. 1998. Summer bed sites of Elk (*Cervus elaphus*) in the Black Hills, South Dakota: considerations for thermal cover management. *The American Midland Naturalist* 139: 133–140.

Moberg, G.P. 1985. Biological response to stress: Key to assessment of animal well-being? *In: Animal Stress.* Moberg, G.P. (Ed.). Baltimore: Waverly Press. pp. 245-267.

Moen, A., Chevlier, S. 1977. Analysis of telemetered ECG signals from white-tailed deer. *Proc. 1st Int. Conf. Wildl. Biotel.* pp. 118-125.

Moen, A., Della Fera, M.A., Hiller, A.L., Buxton, B. 1977. Heart rates of white-tailed deer fawns in response to recorded wolf howls. *Can. J. Zool.* 56: 1207-1210.

Moen, A.N., Whittemore, S., Buxton, B. 1982. Effects of disturbance by snowmobiles on heart rate of captive white-tailed deer. *New York Fish and Game Journal* 29: 176-183.

Morellet, N., Verheyden, H., Angibault, J.-M., Cargnelutti, B., Lourtet, B., Hewison, M.A.J. 2009. The effect of capture on ranging behaviour and activity of the European roe deer *Capreolus capreolus. Wildl. Biol.* 15(3): 278-287.

Mrlik, W. 1990. Disturbance of the roe deer (Capreolus capreolus) in agrocoenoses of southern Moravia. *Folia Zoologica* 39: 25-35.

Müller-Using, D., Schloeth, R. 1967. Das Verhalten der Hirsche (Cervidae) [The behavior of deer], 10. *In: Handbuch der Zoologie 8.* Helmcke, J.-G., Lengerken, H.v., Starck, D., Wermuth, H. (Eds.). Berlin: Walter de Gruyter and Co. 1-60.

Mysterud, A. 1996. Bed-site selection by adult roe deer *Capreolus capreolus* in southern Norway during summer. *Wildl. Biol.* 2 (2): 101-106.

Mysterud, A., Lian, L.-B., Hjermann, D. Ø. 1999. Scale-dependent trade-offs in foraging by European roe deer (Capreolus capreolus) during winter. *Can. J. Zool.* 77: 1486–1493

Mysterud, A., Østbye, E. 1995. Bed-site selection by European roe deer (*Capreolus capreolus*) in southern Norway during winter. *Can. J. Zool.* 73: 924-932.

Naylor, L.M., Wisdom, M.J., Anthony, R.G. 2009. Behavioral Responses of North American Elk to Recreational Activity. *J. Wildl. Manage.* 73 (3): 328–338.

Onderscheka, K. 1979. Untersuchungen der Wechselwirkung zwischen Rotwildpopulation und Umwelt unter besonderer Berücksichtigung der Ursachen der Wildschäden [Research of interactions between red deer populations and the environment under special consideration of game damage]. *Österreichs Weidwerk* 9: 480-485.

Onderscheka, K. 1986. Ist die Fütterung des Rotwildes in der Kulturlandschaft des alpinen Raumes eine biologische Absurdität oder ein Beitrag zur Erhaltung der Funktion des Ökosystems? [Is feeding of red deer in cultural landscape of Alpine regions a biological absurdity or a contribution to sustaining the function of the ecosystem?] P. *CIC-Red Deer Conf. Graz* 1986: 386-395.

Ozoga, J.J., Verme, L.J. 1970. Winter feeding patterns of penned white-tailed deer. *J. Wildl. Manage.* 34: 431-439.

Pagon, N. 2010. Aspect of some ecological characteristics of roe deer (*Capreolus capreolus* L., 1758) population in north-eastern Apennines, Arezzo province, Italy. *Doctoral Thesis.* 159 pp.

Papouchis, C.M., Singer, F.J., Sloan, W.B. 2001. Responses of Desert Bighorn Sheep to Increased Human Recreation. *J. Wildl. Manage*: 65(3): 573-582.

Petrak, M. 1988. Skilanglauf und Rothirsch (*Cervus elaphus* Linné, 1758) in der Eifel. *Z. Jagdwiss.* 34: 105-114.

Petrak, M. 1996. Der Mensch als Störgröße in der Umwelt des Rothirsches (*Cervus elaphus* L. 1758) [Man as a disturbing factor in the environment of the red deer]. *Z. Jagdwiss.* 42: 180-194.

Pfaff, D.W 1982. Motivational concepts: Definitions and distinctions. *In: The physiological mechanisms of motivation.* Pfaff, D.W. (Ed.). New York: Springer. 3-24.

Porges, S.W. 1985. Spontaneous oscillations in heart rate: Potential index of stress. *In: Animal stress*. Moberg, G.P. (Ed.). Baltimore: Waverly Press. pp. 97-111.

Preisler, H.K., Ager, A.A., Wisdom, M.J. 2006. Statistical methods for analysing responses of wildlife to human disturbance. *J. Appl. Ecol.* 43 (1): 164–172.

Prior, R. 1968. The roe deer of Cranborne Chase: *an ecological survey*. London: Oxford University Press. 222 pp.

Putman, R.J. 1988. *The natural history of deer*. Comstock Pub. Associates. 191pp.

Raesfeld, F.v., Neuhaus, A.H., Schaich, W. 1985. *Das Rehwild. Naturgeschichte, Hege und Jagd* [Roe deer. Natural history, management und hunting]. 9.Ed. Hamburg: Paul Parey. 453 pp.

Raesfeld, F.v., Reulecke, K. 1988. *Das Rotwild. Naturgeschichte, Hege, Jagdausübung*. [Red deer. Natural history, management und hunting]. Hamburg: Paul Parey. 416 pp.

Rauer-Gross, B., Gossow, H., Hamr, J., Czabert, H. 1988. Beiträge zur Verhaltensökologie der Gemse (*R. rupricapra r.*) in touristisch stark belasteten alpinen Gebieten [Contributions to behavior ecology of chamois in Alpine regions heavily used by tourists]. *CIC: Gamswildsymposium*. Ljubljana 25./26.10.1988. 127-143.

Reichholf, J., Schemel, H. 1988. Segelsport und Naturschutz - gehört das Segeln aus ökologischer Sicht zur "`ruhigen Erholung"'? [Sailing and nature conservation – is sailing a „quiet recreation" from an ecological viewpoint?] *Zeitschrift für angewandte Umweltforschung* 1(4): 343-354.

Reimoser, F. 1991. Schwerpunktbejagung und Intervallbejagung – Jagdstrategien zur Erhaltung von Wald und Wild [Focus hunting and interval hunting – hunting strategies for maintaining forest and wildlife]. *Österreichs Weidwerk* 12: 35–38.

Reimoser, F. 2005: Freizeitaktivitäten und Wildtiere: Folgen für den Wald [Outdoor recreation and wild animals: Consequences for the forest]. *In: Freizeitaktivitäten im Lebensraum der Alpentiere* [Outdoor recreation in the habitat of Alpine wildlife]. Ingold, P. (Ed.). Bern: Haupt. pp. 311-321.

Reimoser, F., Mayer, H., Holzinger, A., Zandl, J. 1987. Effects of summer and winter tourism on game damage to forests in Angertal, Badgastein, Austria. *Cbl. ges.Forstwesen* 104: 95-118.

Reimoser, F., Ellenberg, H., 1999. Forest management systems as a component of ungulate-game pest management with special reference to roe deer and edge effects. *In: Advances in Vertebrate Pest Management.* Cowan, D.P., Feare, C.J. (Eds.). Fürth: Filander-Verlag, pp. 219–238.

Roshchevskii, M.P., Konovalov, N.I., Beznosikov, V.S. 1976. Cardiac component in emotional stress in elk *Alces alces* and reindeer *Rangifer tarandus* . *J. Evol. Bioch. Physiol.* 12: 347-349.

Rushen, J. 1991. Problems associated with the interpretation of physiological data in the assessment of animal welfare. *Appl. Anim. Behav. Sci.* 28: 381-386.

Rykiel, E.J. 1985. Towards a definition of ecological disturbance. *Aust. J. Ecol.* 10: 361-365.

Schleidt, W.M., Shalter, M.D. 1983. The effect of spatial context on habituation to a predator model. *Z. Tierpsychol.* 61: 67-70.

Schmidt, K. 1992. Über den Einfluss von Fütterung und Jagd auf das Raum-Zeit-Verhalten von alpinen Rotwild [On the influence of feeding and hunting on the space/time behavior of alpine red deer]. *Z. Jagdwiss.* 38: 88-100.

Schober, F., Pasterk, W., Sommer, F. 1982. Implantable long-range, long-life, heart rate and activity transmitter for free-ranging animals. In: Stanford University: P. *7th Int. Symp. Biot. Stanford* 1982. pp. 108-111.

Schober, F., Fluch, G. 1995. Heart rate and body temperature repeater telemetry system for wildlife animals. In: Biotelemetry XIII, P. *13th Int. Symp. Biot.* pp. 366–371.

Schultz, R.D., Bailey, J.A. 1978. Responses of national park elk to human activity. *J. Wildl. Manage.* 42: 91-100.

Selye, H. 1936. A syndrome produced by diverse nocuous agents. *Nature* 138: 32.

Shalter, M.D. 1984. Predator-prey behavior and habituation. *In: Habituation, sensitization and behavior.* Peeke, H.V.S., Petrinovich, L. (Eds.). Orlando: Academic Press. pp. 349-383.

Sibbald, A.M., Hooper, R.J., Gordon I.J., Cumming, S. 2001. Using GPS to study the effect of human disturbance on the behaviour of red deer stags on a highland estate in Scotland. In: *Tracking animals with GPS.* (ISBN 0 7084 0643 2). pp. 39-44.

Sibbald, A.M., Hooper, R.J., McLeod, J.E., Gordon I.J. 2011. Responses of red deer (*Cervus elaphus*) to regular disturbance by hill walkers. *Eur. J. Wildl. Res.* Published online. 9 pp.

Silver, H., Colovos, N.F., Holter, J.B., Hayes, H.H. 1969. Fasting metabolism of white-tailed deer. *J. Wildl. Manage.* 33: 490-498.

Skogland, T., Grøvan, B. 1988. The effects of human disturbance on the activity of wild reindeer in different physical condition. *Rangifer* 8: 11-20.

Smith, E.N., Worth, D.J. 1979. Atropine effect on fear bradykardia of the eastern cottontail rabbit, *Sylvilagus floridanus* . *In: A handbook on biotelemetry and radio-tracking.* Amlaner, C.J., MacDonald, D.W. (Eds.). Oxford: Pergamon Press. pp. 549-556.

Snetsinger, S.D., White, K. 2009. *Recreation and trail impacts on wildlife species of interest in Mount Spokane State Park.* Pacific Biodiversity Institute, Winthrop, Washington. 60 p.

Sparrowe, R.D., Springer, P.F. 1970. Seasonal activity patterns of white-tailed deer in eastern South Dakota. *J. Wildl. Manage.* 34: 420-431.

Squibb, R.C., Kimball, J.F.Jr., Anderson, D.R. 1986. Bimodal distribution of estimated conception dates in rocky mountain elk. *J. Wildl. Manage.* 50: 118-122.

Stadler, R. 1985. Reaktionen des Rehwildes auf Störungen in ballungsnahen Waldgebieten am Beispiel des Nürnberger Reichwaldes [Reaction of roe deer towards disturbances in suburban forest areas by example of the Nürnberger Reichswald forest]. Fachhochschule Weihenstephan. *Fachbereich Forstwirtschaft. Diploma thesis.*

Stankowich, T. 2008. Ungulate flight responses to human disturbance: A review and meta-analysis. *Biol. Conserv.* 141: 2159-2173.

Stemp, R.E. 1983. *Heart rate responses of bighorn sheep to environmental factors and harassment.* Masters Degree Project, Faculty of Environmental Design, Univ. of Calgary. Alberta, Canada.

Sterl, P., Brandenburg, C., Arnberger, A. 2008. Visitors' awareness and assessment of recreational disturbance of wildlife in the Donau-Auen National Park. *J. Nature Conserv.* 16 (3): 135-145.

Stöhr, W. 1982. Telemetrische Langzeituntersuchung der Herzfrequenz von Tupaia belangeri: Basalwerte sowie phasische und tonische Reaktionen auf nichtsoziale und soziale Belastungen [Telemetric long-term investigation of heart rate in Tupaia belangeri]. Universität Bayreuth. Fakultät Biologie, Chemie und Geowissenschaften. *Thesis.*

Stott, G.H. 1981. What is animal stress and how is it measured? *J. Anim. Sci.* 52: 150-153.

Sunde, P., Olesen, C.R., Madsen, T.L., Haugaard, L. 2009. Behavioural Responses of GPS-Collared Female Red Deer Cervus elaphus to Driven *Hunts. Wildl. Biol.* 15(4): 454-460.

Sweeney, J.R., Marchinton, R.L., Sweeney, J.M. 1971. Responses of radio-monitored white-tailed deer chased by hunting dogs. *J. Wildl. Manage.* 35: 707-716.

Taylor, A.R., Knight, R.L. 2003. Wildlife Responses to Recreation and Associated Visitor Perceptions. *Ecological Applications* 13 (4): 951-963.

Thomson, R. F., Spencer, W. A. 1966. Habituation: a model phenomenon for the study of neuronal substrates of behavior. *Psychol. Rev.* 73: 16-43.

Turner, D.C. 1979. An Analysis of Time-Budgeting by Roe Deer (Capreolus capreolus) in an Agricultural Area. *Behaviour* 71: 246-290.

Turner, D.C. 1980. A multi-variate analysis of roe deer (Capreolus capreolus) population activity. *Revue Suisse Zool.* 87: 991-1002.

Ueckermann, E., Goepel, G., Zander, J., Scholz, H., Lülfing, D. 1975. Die Auswirkung der zunehmenden Inanspruchnahme des Waldes durch die erholungsuchende Bevölkerung auf das Verhalten des Wildes und die Bejagungsmöglichkeiten der Wildbestände [The effect of increasing utilization of forests by people seeking recreation on the behavior of wildlife, and hunting opportunities on game populations]. *Z. Jagdwiss.* 21: 50-63.

Vistnes, I, Nellemann, C. 2008. The matter of spatial and temporal scales: a review of reindeer and caribou response to human activity. *Polar Biol.* 31: 399–407.

Völk, F., Gossow, H. 1997: Freizeitaktivitäten und Wildschäden. Schlussfolgerungen aus der wissenschaftlichen und anwendungsorientieren Fachliteratur [Outdoor recreation and game damage. Conclusions of scientific research and practical experience, based on a literature review]. *Cbl. ges. Forstwesen* 114 (1). 35-57.

von der Ohe, C.G., Servheen, C 2002. Measuring stress in mammals using fecal glucocorticoids: opportunities and challenges. *Wildl. Soc. Bull.* 30(4): 1215-1225.

Wallach, A.D., Shanas, U., Inbar, M. 2010. Feeding activity and dietary composition of roe deer at the southern edge of their range. *Eur. J. Wildl. Res.* 56: 1-9.

Walther, F.R. 1969. Flight behaviour and the avoidance of predators in Thomson's gazelle (*Gazella thomsomi* Guenther, 1884). *Behavior* 34: 184-221.

Ward, A.L. 1984. The response of elk and mule deer to firewood gathering on the Medicine Bow Range in southcentral Wyoming. *Proceedings of the 1984 western states and provinces elk workshop*. 28-40.

Ward, A.L.W., Cupal, J.J. 1979. Telemetered heart rate of three elk as affected by activity and human disturbance. *In: Proceedings of the Symposium on Dispersed Recreation and Natural Resource Management: A Focus on Issues, Opportunities and Priorities*. Ward, A.L.W., Cupal, J.J. (Eds.). Utah State University, Logan. pp. 47–55.

Webb, S.L., Gee KI.L., Strickland, B.K., Demarais, S., DeYoung, R.W. 2010. Measuring Fine-Scale White-Tailed Deer Movements and Environmental Influences Using GPS Collars. *Int. J. Ecol.* 12 pp., doi:10.1155/ 2010/459610.

Werner J.-S. 1976. Telemetrische Messungen an Rehen zur Prüfung der Reaktion auf simulierte Umweltreize. [Telemetric measurements on roe deer to survey the reactions towards simulated stimuli]. University of Veterinary Medicine Hannover. *Thesis*.

Whittaker, D., Knight, R.L. 1998. Understanding wildlife responses to humans. *Wildl. Soc. Bull.* 26: 312–317.

Whitten, K.R., Garner, G.W., Mauer, F.J., Harris, R.B. 1992. Productivity and early calf survival in the Porcupine Caribou Herd. *J. Wildl. Manage.* 56: 201-212.

Zhang M. H. and Xiao Q. Z. 1990. A study on feeding and bedding habitat selection by red deer in winter. (Chinese with English abstract) *Acta Theriol. Sinica* 10: 175–183.

Zimmer, H., Vossel, G., Fröhlich, W.D. 1989. Antizipatorische und ereignisbezogene Veränderungen der Schlagfrequenz des Herzens als Indikatoren der Aufmerksamkeitsregulation [Anticipatory and event related changes in heart rate as indicators of attention processes]. *Arch. Psychol.* 141: 251-272.

In: Deer: Habitat, Behavior and Conservation ISBN 978-1-62100-676-3
Editors: A. A. Cahler and J. P. Marsten

Chapter 2

DEER AS BIOMONITORS OF RADIOACTIVE CONTAMINATION

J. Guillén, A. Baeza, A. Salas, and J. A. Corbacho

LARUEX (Environmental Radioactivity Laboratory), Dpt. Applied Physics, Faculty of Veterinary Sciences, University of Extremadura, Avda. Universidad, s/n, 10003, Cáceres, Spain

ABSTRACT

Adequate radiological protection of the environment requires the identification of animals sensitive to potential variations of the radionuclide content in their surroundings. Deer are considered suitable candidates for biomonitors in forest ecosystems. Due to their wide geographical distribution and their varied diet, the ICRP (International Commission on Radiological Protection) considers deer a reference animal. Deer also constitute a significant pathway for the transfer of radionuclides to man, especially in areas where a major deposition of anthropogenic radionuclides, such as $^{134,137}Cs$ and ^{90}Sr, occurred. The uptake of radionuclides takes place via the deer's diet. The radionuclide content in the plants they eat depends on various factors: plant species, soil properties, and the amount of the deposition, among others. The consumption of mushrooms is also an important source of radionuclides in deer. Depending on their chemical properties, these radionuclides are accumulated in different tissues of the animal's body. Radiocæsium is a chemical analogue of potassium, and is located preferentially in muscle tissue, and hence is important in the deer–man pathway. Also, naturally occurring radionuclides of the uranium series, such as ^{210}Pb and ^{210}Po, accumulate in muscle. Radiostrontium has a different chemical behaviour, as it is a chemical analogue of calcium. The ^{90}Sr is mainly detected in calcified tissues, such as bones and antlers. Because the antlers are shed annually, they constitute an excellent biomonitor for the evaluation of the temporal and spatial variations of radiostrontium in the environment. However, some factors, such as the age of the animal or its health status, may affect calcium remobilization, and hence cause certain variations in the radiostrontium content of the antler.

INTRODUCTION

Mammals are considered useful organisms for biomonitoring purposes because they can be used to obtain temporal and spatial information on an environmental pollutant. Especially interesting, therefore, are free-range or wild animals. They consume local flora and fauna that also reflect the environmental quality status of the soil, water, and air which might be affected by contamination. Such contamination will affect the animals and be reflected in their health. The criteria for appropriately selecting animal organisms for biomonitoring purposes are [1, 2]: spatial representativeness; ecological representativeness; availability; the practicability of specimen collection and the experimental suitability; trophic level within an ecosystem; and reproducibility of specimen collection.

The members of the family *Cervidae*, which includes reindeer, moose, elk, red deer, caribou, etc., are widespread throughout the world. Their body size is large enough to provide adequate tissue samples for radionuclide determinations. Usually they are hunted, so that no additional killing is needed to provide samples. In the case of antlers, since they are usually shed and re-grown annually, they can be collected without disturbing the animal. These animals also eat a wide variety of the available vegetation, and their stomachs can be considered a complete "botanical inventory of the area" in which they live [3]. Thus, they can be regarded as good "samplers" of the area. However, the fact that the winter and the summer ranges differ may present some limitations, since the animals may migrate over many kilometres to obtain food supplies. This migration is limited in many areas of Central and Southern Europe due to the presence of anthropogenic barriers such as roads, human settlements, and private hunting preserves [2, 4]. As humans are mammals too, the results of monitoring them radiologically have a high degree of validity for the protection of human health.

Traditionally, radiological protection of the environment has been viewed anthropocentrically, assuming that if one ensures that humans are protected against ionizing radiation, it will follow that other living beings will also be adequately protected [5]. From this point of view, knowledge of the radionuclide content in deer is necessary because they are usually hunted for food [6], and the ingestion of their radioactive content by humans can represent a significant contribution to the internal dose, especially for certain human population groups in the Arctic [7, 8]. However, the notion that if humans are properly protected then other species will not be put at risk is no longer tenable. Exposure scenarios and pathways for human protection may not apply to non-human biota. To deal with this problem, a set of Reference Animals and Plants were selected in a similar way to the protocols for human radiological protection [9, 10]. These Reference Animals and Plants, which include deer, can be regarded as hypothetical entities with certain assumed basic biological characteristics of a particular type of animal or plant, describing the generality of the taxonomic level of Family, with specific anatomical, physiological, and life-history characteristics.

DETERMINATION OF RADIONUCLIDES

Radionuclides are unstable isotopes that usually lose their excess energy by the emission of ionizing radiation. The emission probability is different for each radionuclide and is expressed in terms of its half-life. The half-life of a radionuclide, $T_{½}$, is the time it takes for the activity to decrease by a half. There are three main types of ionizing radiation: α, β, and γ, whose main properties are the following:

1. α-particles are helium nuclei (2 protons + 2 neutrons). They have high values of mass and kinetic energy, within the range 3.95 – 8.78 MeV[1]. Their interaction with matter is so strong that even a sheet of paper is enough to stop them, with all their energy being lost in crossing its

[1] An electron volt (eV) is a unit of energy equal to the amount of kinetic energy gained by a single unbound electron when it accelerates through an electric potential difference of 1 V. 1 eV = $1.609 \cdot 10^{-9}$ J.

thickness. Some α-emitting radionuclides are: ^{238}U, ^{226}Ra, ^{210}Po, ^{232}Th, etc.

2. β-particles have various origins depending on the characteristics of the nucleus (β^-, β^+, EC or electron conversion, ...), but the final result is the emission of electrons or positrons within a wide range of energies (0.01859 -3.54 MeV). Due to their low mass and charge compared with α-particles, they can penetrate deeper into matter, requiring a few centimetres of aluminium to stop them. Some β-radionuclides are: ^{3}H, ^{210}Pb, ^{90}Sr, etc.
3. γ-radiation consists of high-energy photons, from a few keV to less than 10 MeV. Their penetrating power is greater than the previous particles, and they require several centimetres of lead to be stopped. ^{40}K, ^{137}Cs, among others, are γ-emitting radionuclides.

Systems to detect radionuclides are different for each type of radiation. Due to the low penetrating power of α and β particles, it is necessary to perform radiochemical separation procedures in order to determine them properly. In the case of γ-radiation, however, such radiochemical separations are usually unnecessary. The determination of α-emitting radionuclides is usually by means of silicon detectors within a vacuum chamber. Gas-flow proportional counters and LSC (liquid scintillation counting) are used to determine both β-particles and α-particles. The determination of γ-emitting radionuclides is by means of NaI(Tl) and HP(Ge) crystals. Sample preparation for γ-spectrometry is easier than for α- and/or β-spectrometry because usually no radiochemical procedures are involved. As an example, the preparation of the sample for the determination of ^{137}Cs in meat samples can simply consist of reducing its volume by drying and then ashing at 400 °C.

SOURCES OF RADIONUCLIDES IN THE ENVIRONMENT

The radionuclides occurring in the environment can be classified into two main groups: natural and anthropogenic (or man-made). The naturally occurring radionuclides comprise the natural radioactive series of radionuclides, ^{40}K, and cosmogenic radionuclides. Cosmogenic radionuclides are formed by the interaction of high-energy cosmic rays with atomic nuclei in the atmosphere. Some examples are ^{7}Be, ^{10}Be, and ^{22}Na. There are three natural radioactive series present in nature: thorium, uranium, and actinium, also denominated A = 4n, A = 4n+2, and A = 4n+3 respectively, where A is the mass number of the radionuclide. These natural series begin with ^{232}Th (thorium series), ^{238}U (uranium series), and ^{235}U (actinium series), and finish in ^{208}Pb, ^{206}Pb, and ^{207}Pb, respectively. The isotope ^{40}K ($T_{½}$ = $1.277 \cdot 10^9$ yr) is also a naturally occurring radionuclide with a long half-life, and constitutes 0.012% of potassium. The distribution of these radionuclides in the geosphere depends on the geological media from which they derive, and the processes which concentrate them at a specific location in specific media [11]. They can be enhanced locally by human activities, and may be a cause for significant radiological concern because of NORM (Naturally Occurring Radioactive Material) industries. These industries use raw materials with high contents of these radionuclides, and their bioavailability in their products, by-products, and residues may be enhanced due to physicochemical changes or to the form in which they are managed. Mining industries (uranium, iron, copper, ...), the production of fertilizers from phosphate rocks, and burning non-nuclear fuels (coal, oil, gas) are some examples of NORM activities [11].

The anthropogenic radionuclides were first introduced into the environment in 1945 as a consequence of the atom bomb blasts during World War II. During the 1950s and 1960s, there were a great number of atmospheric nuclear weapons tests, which released huge amounts of a great number of radionuclides into the atmosphere. As a consequence of atmospheric circulation, they became worldwide distributed and ultimately were deposited onto the soil [12]. This phenomenon is usually known as global fallout. Most of these radionuclides were short-lived and have already decayed, as is the case with ^{131}I ($T_{½}$ = 8.02 d). However, ^{137}Cs and ^{90}Sr have longer half-lives (30.07 and 28.79 yr respectively), and are the most important man-made radionuclides environmentally. The estimated quantities of activity released were 948 and 622 PBq for ^{137}Cs and ^{90}Sr, respectively. Some radioisotopes of plutonium were also released but in lower amounts – 10.87 PBq for $^{239+240}Pu$. The annual deposition of ^{137}Cs and ^{90}Sr due to global fallout is shown in Figure 1. There was a maximum around 1962, in which year the greatest number of tests were carried out. As the majority of tests took place in the northern hemisphere, the deposition there was greater than in the southern hemisphere. The deposition was inhomogeneous in latitude because of circulation patterns and airflows in the stratosphere and troposphere. It was greater in the temperate regions, with a maximum deposition of ^{90}Sr in the 40–50° latitude bands, and lower in the Equatorial and Polar regions [12].

There is another natural series named neptunium (A = 4n+1), which was extinct because the half-life of ^{237}Np ($T_{½}$ = $2.14 \cdot 10^6$ yr) is shorter than the

Earth's age. However, it was re-introduced into nature by the atmospheric release of anthropogenic ^{241}Pu and ^{241}Am, predecessors of ^{237}Np.

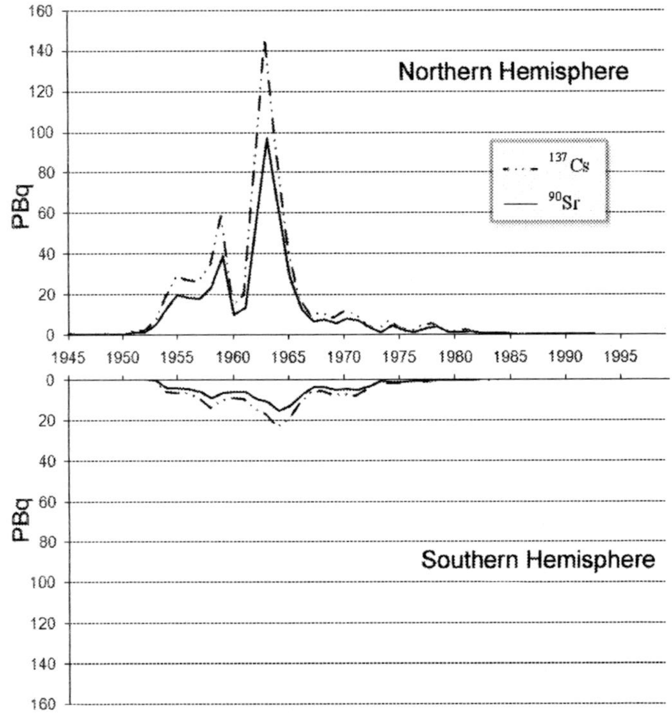

Figure 1. Annual deposition of ^{137}Cs and ^{90}Sr, expressed in petabecquerels (PBq), in the northern and southern hemispheres produced by atmospheric nuclear tests. Data from UNSCEAR [12].

Other sources of radionuclides in the environment are related to releases from nuclear fuel reprocessing factories, plutonium fabrication plants, nuclear waste storage sites, and accidents in nuclear reactors such as Windscale, Three Mile Island, Chernobyl, and, more recently, Fukushima. The most serious accident was Chernobyl, which marked a point of inflexion for radioecology and radioprotection. Large quantities of radionuclides were released, about 54, 85, and 10 PBq for ^{134}Cs, ^{137}Cs, and ^{90}Sr, respectively [12]. The most heavily contaminated areas (>3700 kBq/m^2 of ^{137}Cs) were located mostly within the 30 km radius around Chernobyl. However, it also seriously affected other countries such as Finland, Sweden, Norway, Germany, and Austria, where there were areas in which about 40-185 kBq/m^2 of ^{137}Cs were deposited [12].

RADIONULIDE CONTENT IN DEER

There are many articles in the literature dealing with the radionuclide content of different parts of various deer species. Table 1 presents the ranges of variation for several anthropogenic (^{137}Cs, ^{90}Sr, $^{239+240}$Pu, ^{214}Am) and naturally occurring (U_{nat}, ^{226}Ra, ^{210}Pb, ^{210}Po) radionuclides in different organs. The radionuclide accumulation is different for each organ since it depends on the element's chemical characteristics.

Naturally Occuring Radionuclides

The ^{40}K content in liver, kidney, and muscle of deer is very homogeneous, and usually greater than the bone content [13-16]. The maximum value of the range in bones presented in Table 1 is due to a single high value detected in the Russian Arctic for reindeer [15]. No comparison between muscle and bone content was possible in that study, because it only analyzed calcified tissues. It has also been reported that the ^{40}K content in liver, kidneys, and muscle is similar among different herds foraging in different areas [16], and it has been proposed that these parameters could be used as quality control measurements between different studies [13]. This similarity between different herds was to be expected, especially in areas not affected by NORM industries, because ^{40}K is 0.012% of the potassium present in the environment, and potassium is an essential element for life.

Deer are also able to accumulate other naturally occurring radionuclides, such as uranium, radium, ^{210}Pb, and ^{210}Po. These radionuclides may be a cause for radiological concern if associated with NORM industries in the area. They have been analyzed in different tissues of caribou (*Rangifer tarandus*) near uranium mining installations [14], detecting a mean value of 0.040 Bq/kg (1.6 ppb) in muscle for natural uranium, U_{nat}. This parameter, U_{nat}, is the sum of activities due to ^{234}U ($T_{½} = 2.455 \cdot 10^5$ yr), ^{235}U ($T_{½} = 7.038 \cdot 10^8$ yr), and ^{238}U ($T_{½} = 4.468 \cdot 10^9$ yr). Muscle tissues had similar contents to the liver, but less than kidney or bones (see Table 1). The uranium content in muscle is similar for mule deer (*Odocoileus hemionus*) and elk (*Cervus elaphus*) tissues [6]. The excretion of uranium in fæces shows higher specific activity, about 1.9 Bq/kg [14], but any interpretation of this difference would require the differences in body and fæces weight to be taken into account. The ^{226}Ra($T_{½}$ = 1600 yr) content presents a similar pattern to that of uranium, although the content in the liver is slightly greater than in muscle tissue, and both are less than the kidney content [14]. The main difference from uranium is due to radium's chemical behaviour. Radium is an alkaline earth, as is calcium. It therefore mainly accumulates in calcified tissue, and its content in bone, including antlers, is greater than in any other tissue [14, 15].

Knowledge of the content of ^{210}Pb and ^{210}Po in different tissues is very important because of their great radiotoxicity and major contribution to the internal dose by ingestion. These two radionuclides are descendants of ^{226}Ra, and therefore closely related. The ^{210}Pb ($T_{½}$ = 22.3 yr) decays to ^{210}Bi ($T_{½}$ = 5.013 d) and also to ^{210}Po ($T_{½}$ = 138.376 d). The ^{226}Ra content in bones has been observed to be less than the ^{210}Pb content [15], showing that there had to be another source for ^{210}Pb in addition to the decay of ^{226}Ra. This source is the consumption of lichens [15, 17]. The ^{210}Pb and ^{210}Po content in different tissues increases in the following order: muscle < kidney < liver < bone [14, 16-18]. Although the two radionuclides are related, their content in different tissues is not the same, implying that they are not in secular equilibrium[2]. The ^{210}Po/^{210}Pb ratio differs from one tissue type to another. In muscle, liver, and kidney, the ratio is greater than unity [14, 16, 18], implying a greater content of ^{210}Po than ^{210}Pb. The ^{210}Pb content in bone and the ratio ^{210}Po/^{210}Pb in muscle and liver tissues are uncorrelated with age [14, 15, 17].

Anthropogenic Radionuclides

Although the values of $^{239+240}$Pu and ^{241}Am fallout are less than those of other man-made radionuclides, they may be enhanced in the vicinity of potential source terms, such as radioactive waste disposal sites. However, an analysis of the tissues of mule deer and elk in such a location [6] showed their activity levels to be the lowest reported for anthropogenic and naturally occurring radionuclides (see Table 1), implying that there was no enhancement due to the waste disposal site. There were no differences between the two species of deer.

Radiostrontium is a chemical analogue of calcium – both are alkaline earth elements. It is therefore accumulated preferentially in bones, as is calcium (see Table 1). Different species of *Cervidae* present similar ranges of ^{90}Sr content in bones [19]: *Cervus elaphus* (71-899 Bq/kg), *Capreolus capreolus* (84-437

[2] Secular equilibrium is a situation in which the quantity of a radioactive isotope remains constant because its production rate, due to decay of a parent isotope, is equal to its decay rate.

Bq/kg), and *Alces alces* (105-667 Bq/kg)[3]. This suggested that these species accumulate ^{90}Sr in bone in approximately the same way, reflecting the ^{90}Sr deposited on soil. Observations of the variation of the ^{90}Sr content in *Rangifer tarandus* bones collected at the Novaya Zemlya test site in Russia confirm this suggestion, with the content in bones of animals that lived during the period of underground nuclear testing being about 750 Bq/kg, whereas the content after the tests had been banned was about 440 Bq/kg [15]. A correlation has been observed between ^{90}Sr bone content and site altitude [21], reflecting the relationship between altitude and rainfall, and hence the ^{90}Sr deposition [22]. Stable strontium also accumulates in bone, but there is no observed correlation with ^{90}Sr [19]. No seasonal trend of ^{90}Sr content in bones has been observed [17, 20]. The long-term accumulation of ^{90}Sr decreases with time, with an ecological half-life of about 9 yr [17]. The relationship between age and ^{90}Sr bone content is unclear. Some authors detect greater ^{90}Sr content in bones of old (about 10 years in age) females than in calves [17, 21], with this being attributed to the cumulative deposit of ^{90}Sr in bone tissue. However, the transfer of dietary strontium and calcium to bones is reduced in older animals [23], so that one would expect a negative correlation.

Antlers are a special case of calcified tissue. They grow rapidly – those of red deer (*Cervus elaphus*) by on average (1.95 ± 0.05) cm per week in the males' first season [24]. They end by constituting from 1% to 5% of the animal's body weight [25]. This growth implies a significant nutritional stress, which is provided from food intake rather than remobilization from bone tissue [25, 26]. Since antlers are shed and re-grown annually, the time period during which any strontium was incorporated is clearly defined. Another advantage is that antlers are shed naturally, and may be collected without disturbing the animal. Hence, the concentrations of "bone-seeking" radionuclides in hard antlers constitute cumulative measures of the exposure of the forming antler during a well-defined period [2].

The content of ^{90}Sr in antlers has been used as a biomonitor of radiostrontium in an ecosystem. It reflects the geographical deposition of ^{90}Sr in the soil from different sources [4, 26-28]. No increase in the ^{90}Sr signal was observed in antlers from Austria or Denmark after the Chernobyl accident due to the lower deposition of ^{90}Sr from that source [26, 28]. The ^{90}Sr is homogeneously distributed within the antler [4], as also is the calcium content [25]. The calcium and magnesium contents in antlers have been shown to be

[3] The original data were given as Bq/kg-ash, converted to Bq/kg using the factor 0.7 kg ash/kg d.w. from the literature [20].

correlated, as also have the stable strontium and ^{90}Sr contents, but there is no correlation between ^{90}Sr and ^{226}Ra [4]. The lack of correlation of strontium with calcium or magnesium reflects the irregularity in its incorporation via the animal's diet [29] due to seasonality in the selection deer make in their browsing during the summer–autumn period. It seems that some plant taxa might represent more effective routes for heavy metal uptake by mammals than others. The ^{90}Sr ratio between antlers and bones is about unity for calves [17]. It reached a value of 2.3 in Denmark during 1989, which was attributed to enhanced ^{90}Sr content in the food supply, and fell back to about unity in the period 1990–92 [26]. This confirmed that the ^{90}Sr content in the antlers came mainly from the food intake. The ^{90}Sr content in antlers decreases with time, showing an ecological half-life of 5.86 yr. A negative correlation has been observed between age and ^{90}Sr in antlers [27].

Radiocæsium is the most extensively studied radionuclide in deer species because of its environmental significance. It presents the greatest content in muscle tissue, followed by kidneys. Similar ^{137}Cs contents have been observed in muscles of mule deer (*Odocoilus hemionus*) and elk (*Cervus elaphus*) in the same ecosystems, while the values in heart, liver, lung, and rumen tissues were similar to each other but lower than the muscle and kidney tissues [6]. The lowest values of ^{137}Cs are detected in bone and blood [14, 30]. High values have been detected in liver [31], which the authors attributed to differences in soil characteristics, but unfortunately they reported no data on muscle tissue.

The fact that ^{137}Cs is accumulated preferentially in muscle tissues is a serious health concern, especially for some critical populations in the Arctic [13]. There has even been a method developed to estimate caribou consumption from the relationship between this consumption and humans' body burden of ^{137}Cs [32]. However, there is a wide range of variation (about seven orders of magnitude) of ^{137}Cs in muscle tissue (see Table 1). This clearly reflects the quantity of ^{137}Cs deposited in soils. The correlation between ground deposition and ^{137}Cs in muscle tissue has been observed directly or indirectly by various workers [33, 34]. It is clearly appreciated in the data of Table 2 which presents the ranges of variation of ^{137}Cs in muscle for species of *Cervidae* in different countries, considering deposition due mainly to global fallout (before 1986) and the additional contribution from the Chernobyl accident (after 1986). The range of activity levels for ^{137}Cs increased in those countries most seriously affected by this event [12]. In other territories, such as Canada and Greenland, a second maximum appeared, although lower than the first due to global fallout, or else the activity levels simply decreased. After Chernobyl, ^{134}Cs ($T_{½}$ = 2.06 yr) was also detected in muscle [13]. The contribution of the Chernobyl accident was estimated to be 10-40% of the total ^{137}Cs in Canada, Alaska, and Greenland [13]. This estimate was based on the fact that the isotopic ratio $^{134}Cs/^{137}Cs$ occurring in a radionuclide release are well established, and depend on the characteristics of the source (age of the fuel, activity levels present, etc.).

The distribution of ^{137}Cs in muscle tissue is quite homogeneous. Analysis of the different muscle tissues of a 1.5 year old male weighing 66 kg showed that the ^{137}Cs in different muscles did not differ by more than 10% [35]. Young animals showed greater radiocæsium contents than adults due to their higher metabolic rates [2, 36]. This occurred mainly in the first years after Chernobyl in adult animals born before the accident. In the period 2000-2004, no significant differences were reported between calves and adults [37].

The radiocæsium content in muscle has a strong seasonal trend. It reaches maximum values in autumn-winter and minima in spring-summer [2, 13, 38, 39]. This seasonality is mainly due to changes in the availability of forage plants and to the daily ^{137}Cs intake associated with them, and to the animals' nutritional requirements. During summer, the intake of potassium by *Rangifer tarandus* is greatest, and therefore the ^{137}Cs concentration decreases [40]. In Arctic areas, lichen intake in winter represents about 58-67% of the total diet, while in summer vascular plants predominate, representing 82-89% [41]. This variation is confirmed by analyses of the rumen content [36]. Table 3 gives the ranges of variation of ^{137}Cs content in typical diets. While lichens, mosses, and brush can have high concentrations of ^{137}Cs, the ^{137}Cs content of fungi is the greatest. The major seasonality of the appearance of fruiting bodies of fungi in forest ecosystems, with a first maximum in autumn and a second in spring, may explain the high values of radiocæsium in muscle. The production of fungi in autumn depends on the rainfall in July, August and September. A greater content of ^{137}Cs in roe deer has been reported in years when fungi fruiting bodies are abundant [38]. Mushrooms bitten by roe deer have been observed during sample collection, and analyses showed them to have about 33 500 Bq/kg d.w. of ^{137}Cs [42]. A mushroom rumen content of up to 20% in an individual moose has been reported [36]. Abundant spores of the fungi *Xerocomus* spp. and *Hypholoma capnoides* and an increased ^{137}Cs content have been observed in the fæces of roe deer [43]. In Iceland, no significant seasonal variation was observed due to the absence of significant amounts of highly contaminated lichens or mushrooms [44].

The temporal evolution of ^{137}Cs in muscle after the Chernobyl fallout can be described as a combination of two components: an exponential decay, and a peak in the second half of each year due to the mushroom season [39]. The exponential decay is governed by the effective half-life, $T_{½\,eff}$, given by Eq. 1.

$$\frac{1}{T_{1/2\ eff}} = \frac{1}{T_{1/2\ phys}} + \frac{1}{T_{1/2\ ecol}} \qquad (1)$$

where $T_{½\,phys}$ represents the physical half-life of the radionuclide and $T_{½\,ecol}$ the ecological half-life in the ecosystem being considered. In Arctic ecosystems, the effective half-life of ^{137}Cs is in the range 6–20 yr [13, 38, 41, 45]. Changes in the effective half-life have been reported: the ^{137}Cs in reindeer declined rapidly during the first years after the Chernobyl accident, with $T_{½\,eff}$ = 3–5 yr, and slowly in later years $T_{½\,eff}$ = 7–16.8 yr [46]; and in some areas of Norway during the period 1995–2004, there was no decline at all of ^{137}Cs in muscle [37].

Table 1. Ranges of variation of ^{137}Cs, ^{90}Sr, $^{239+240}$Pu, ^{241}Am, ^{40}K, U_{nat}, ^{226}Ra, ^{210}Pb, and ^{210}Po, expressed in Bq/kg f.w., detected in different species of the *Cervidae* family in the literature. U_{nat} is the sum of ^{234,235,238}U activities

Radionuclide	Muscle	Kidney	Liver	Bone	Antler	References
^{137}Cs	0.14-47000	4.32-20300	11-5750	5-820	----	[6, 15, 16, 20, 31, 34, 37, 47]
^{90}Sr	0.006-2.5	----	----	21-1814	6-2400	[4, 6, 15, 19, 20, 21, 27, 28]
$^{239+240}$Pu	0.00015-0.0029	----	----	0.51-3.5	----	[6, 15]
^{241}Am	0.00022-0.001	----	----	----	----	[6]
^{40}K	72.9-167	67-147	51.5-169	40-550	----	[13-16]
U_{nat}	0.0021-0.0073	0.34-0.54	0.049-0.065	0.56-0,78	----	[6, 15]
^{226}Ra	0.11-0.28	1.0-1.4	0.15-2.3	0.5-160	8-36	[4, 14, 15, 18, 48]
^{210}Pb	0.7-250	30.8-185	0.74-791	18-3800	----	[14-18]
^{210}Po	1.5-65	101-504	3.7-1734	11.8-3070	----	[14-18]

Table 2. Radiocæsium content in muscles of different species of the *Cervidae* family, expressed in Bq/kg f.w., before and after 1986

Country	Before 1986	After 1986	References
Canada	24-6420	4-1440	[8, 13, 16, 32, 49, 50]
USA (Alaska)	15-1800	----	[13, 40]
USA	----	0.14-3.2	[6]
Denmark (Greenland)	7-1790	13-331	[13]
Denmark	----	50-1200	[43]
Iceland	----	0.2-22	[44]
Ireland	----	191-367	[34]
UK	----	72-216	[51]
Norway	150-4000	6200 - 20000	[37, 40]
Sweden	8-3900	51-47000	[6, 30, 33, 36, 41, 52, 53, 54]
Finland	240-2500	640-9030	[35, 55]
Austria	----	100-9000	[2, 39, 56]
Germany	----	20-8700	[39]
Czech Rep.	----	3.9-523	[57]

Table 3. Ranges of variation of ^{137}Cs, expressed in Bq/kg d.w., in *Cervidae* diet in Scandinavian countries [36, 41, 42, 55, 58]

Diet	Range ^{137}Cs (Bq/kg d.w.)	Diet	Range ^{137}Cs (Bq/kg d.w.)
Trees	14 - 2500	Bush	43 - 49010
Lichen/Moss	200 – 42180	Forbs	350 - 3800
Grass	455 - 2500	Fungi	20 - 950000

TRANSFER OF RADIONUCLIDES TO DEER

The transfer of radionuclides to *Cervidae* species is governed by various factors. Fundamentally, anthropogenic radionuclides are deposited onto soils, and then taken up by plants which are ingested as food by animals. Therefore, the first factor affecting the transfer is the quantity of radionuclides deposited on the soil, as was discussed in the previous section. Once deposited on the soil, the radionuclides begin to interact with its components and the compounds present in it. One result is that only a certain fraction of the radionuclides present in a soil are available for transfer to plants. This is the case both for anthropogenic and for naturally occurring radionuclides. Soil characteristics such as the pH, the clay and organic matter contents, and the climate conditions to which the ecosystem is subject influence this transfer. For example, radiocæsium is more mobile in acidic soils [22, 59]. A low potassium content and low pH enhances plants' uptake of radiocæsium [60]. The clay content can reduce the availability of radiocæsium, since it tends to be irreversibly bonded to clay particles at the frayed edge sites (FES) [61]. The presence of humic and fulvic acids in a soil is also important: they can form complexes with plutonium and americium [62], and can reduce the adsorption of ^{137}Cs by clays [63]. In temperate environments, a high content of organic matter in forest soils is accompanied by high contents of humic and fulvic acids [64].

The transfer of radionuclides to animals is quantified in terms of the so-called transfer factors. There are several definitions, depending on the perspective being taken. For the daily intake of radionuclides, one uses the fractional transfer factor, F_m (see Eq. 2), defined as the ratio between the radionuclide content in deer, C_{deer}, and the daily intake of radionuclides, I. The usual definition of the transfer factor, TF, or concentration ratio, CR (see Eq. 3), is defined as the ratio between the deer, C_{deer}, and the soil contents, C_{soil}. Finally, for deposition events, one uses the aggregate transfer factor, T_{agg} (see Eq. 4), defined as the ratio between the content in deer and the soil deposition.

$$F_m = \frac{C_{deer}(Bq/kg)}{I(Bq/d)} \tag{2}$$

$$TF, CR = \frac{C_{deer}(Bq/kg\ f.w.)}{C_{soil}(Bq/kg)} \tag{3}$$

$$T_{agg} = \frac{C_{deer}(Bq/kg\ f.w.)}{Soil\ Deposition(Bq/m^2)} \tag{4}$$

The values of F_m for ^{137}Cs in *Cervidae* are in the range 0.24–0.65 d/kg [41, 65]. UNSCEAR has proposed nominal values of transfer factors, TF or CR, for diverse reference organisms [64]. The recommended values for deer of Cs and Sr are 3 and 2, respectively. The aggregate transfer factor is normally used to analyze the transfer to *Cervidae*. The values of T_{agg} for ^{137}Cs are in the range 0.0048–1.8 m^2/kg [46, 57, 64, 66-68]. Different T_{agg} values for ^{137}Cs are obtained in areas with little Chernobyl fallout (0.15–0.32 m^2/kg) than in more heavily contaminated areas (0.78–1.87 m^2/kg) [46]. Also, the T_{agg} values for radiocæsium are lower when the animals have access to farmland, due to the normally high potassium content, pH values, and clay content of their soils [52, 60].

Remediation

After the Chernobyl accident, various remediation procedures were mainly developed and implemented in order to reduce the ^{137}Cs content in meat, as this was a significant radiological hazard for humans. They fall into the following four groups of techniques:

- Altering the slaughter periods: The usual period of slaughter was from mid-November through December, coinciding with the increase of ^{137}Cs content due to mushroom consumption. Earlier slaughter can reduce the ^{137}Cs content because of less time being available for the animals to browse on mushrooms. The effectiveness of this countermeasure depends on accurately predicting the most suitable time [7].
- Uncontaminated feed: Providing the animals with uncontaminated feed for 8-12 weeks before slaughter is capable of reducing the ^{137}Cs to 5-15 % of the initial values [7, 30].
- Addition of cæsium binders: Clay minerals such as bentonite, and iron compounds such as ammonium-ferric(III)-hexacyanoferrate(II) (AFCF), can act as cæsium binders, reducing the absorption of ^{137}Cs. Both options are effective by enhancing the excretion of the ^{137}Cs content with the fæces. The administration of AFCF reduces urinary excretion, however, but it seems to have fewer adverse side effects than bentonite [30, 69, 70].
- Addition of potassium: Since cæsium is a chemical analogue of potassium, the addition of potassium to the diet means that less cæsium is taken up to fill any potassium deficit that might exist. Laboratory experiments showed the addition of KCl to reduce ^{134}Cs absorption in the gut [71].

These remediation techniques were highly effective in the first year after the Chernobyl accident, reducing the dose by 98% [7].

RADIATION DOSE TO DEER

Estimation of the absorbed dose[4] rate in animals and plants is based on the dose conversion coefficients (DCCs) calculated for external and internal exposures to natural and artificial radionuclides of reference organisms. The International Commission on Radiological Protection (ICRP) defines the reference organisms as ‘‘entities that provide a basis for the estimation of radiation dose rate to a range of organisms which are typical, or representative, of a contaminated environment’’ [10]. This publication includes a small set of Reference Animals and Plants published in an earlier work [9], plus their relevant databases, for a few types of organisms that are typical of the major environments. These reference animals can be considered hypothetical entities with basic biological characteristics of a particular type of animal that describes the generality of the taxonomic level of Family. The reference organisms do not correspond to actual individuals or species, but consist of three dimensional ellipsoidal "phantoms" used to calculate internal and external dose rates per unit activity concentration in animals and plants of a given size. Much of the literature data on doses in reference animals are based on mathematical calculations with Monte Carlo models such as ERICA [72].

Annex D of ICRP publication 108 [73] includes a detailed explanation of experimental data on reference animals based on mortality, morbidity, reduced reproductive success, and chromosomal damage and mutations. There have been few studies of the mortality of large mammals as a result of radiation exposure. The only relevant data are those relating to reindeer inhabiting the Novaya Zemlya archipelago during nuclear tests, for which it was estimated that a dose of 8.7 Gy from mixed radiation produced 50% mortality in young animals during the summer and autumn months [4]. For the case of chronic irradiation, there have been some field observations on deer. For example, different species of deer living in a territory contaminated after the Kyshtym accident in 1957 were estimated to have received chronic irradiation from ^{90}Sr (3.7–37 MBq/m^2 in soil). The estimated chronic exposure of European roe

[4] Absorbed dose is the amount of energy that ionizing radiation imparts to a given mass of matter. In other words, the dose is the amount of radiation absorbed by an object. The SI unit for absorbed dose is the gray (Gy).

deer to 0.1–1 Gy/day (cumulative dose of 10– 30 Gy to the intestines) in the autumn of 1957 was followed by decreased numbers in the autumns of 1957 and 1958 in comparison with the period before the accident [75, 76], and estimated accumulated doses of 0.1–1.1 Gy/year in bones were followed by a five-fold decrease in the number of roe deer per km^2 from 1970 to 1977. Calculations of the external dose rate in reference animals using the ERICA tool gave, for roe deer as the reference animal, a range of background absorbed dose rates in England and Wales of 55–87 nGy/h [72]. An evaluation of the regional variation in radionuclide concentration and background radiation dose in caribou (*Rangifer tarandus*) in the Canadian Arctic during 1992–94 [16], found the greatest individual weighted absorbed dose to be in bone (810 mGy/yr) and liver (530 mGy/yr).

There are no morbidity data on deer, and most of the large mammal studies (on dogs and other 'laboratory animals') have focused on providing data for use in developing radiological protection criteria for humans. It is likely, however, that the same general conclusions with regard to the induction of 'tissue reactions' in humans, and possibly also the risk of induction of cancers in relation to dose, also apply more or less equally to other large long-lived mammals. Finally, there have been very few studies on the reproductive capacity of large mammals – in particular, there are no direct studies of deer in this sense. Some of the more complete data sets on 'large' mammals are those relating to pigs. However, human data relating to the sterility of males and females probably apply equally well to other large mammals such as deer.

Conclusion

The *Cervidae* have great importance as bioindicators in the radioprotection of the environment and of humans. The ICRP considers deer to be a reference organism in forest ecosystems. They can accumulate anthropogenic and naturally occurring radionuclides in different organs according to the elements' chemical properties, thus reflecting the radiological status of the environment in which they live. With respect to human radioprotection, radiocæsium is radiologically the most important because it is accumulated preferentially in muscle tissue, and its intake can imply a significant increase in the overall dose received by the population. The ^{137}Cs content in deer muscle is governed by its content in their diet, especially in lichens and mushrooms. In countries heavily contaminated by the Chernobyl accident, the content in deer muscle tissue shows a strong seasonal trend, with a maximum in autumn due to mushroom consumption. Antlers also play an important role as bioindicators, especially for "bone-seeking" radionuclides (^{90}Sr, ^{226}Ra, ^{210}Pb, and ^{210}Po). Because they are shed annually, the sampling procedure does not disturb the animals, and they provide clear information on temporal variations. However, the relationship between the age of the *Cervidae* individuals and radionuclide accumulation in their bone and antlers needs further investigation.

Acknowledgement

We are grateful to the Autonomous Government of Extremadura (Junta de Extremadura) for the financial support to the LARUEX research group (FQM001).

References

[1] Ellenberg, H. In: Ellenberg, H.; Arndt, U.; Bretthauer, R.; Ruthsatz, B.; Steubing, L. (Eds.); *Biological Monitoring: Signals from the Environment;* Vieweg, Braunsweig, 1991, 13-74.

[2] Tataruch, F.; Kiedorf, H. In: Markert, B.A.; Breure, A.M.; Zechmeister, H.G. (Ed.); *Bioindicators and biomonit*ors; Elsevier Science Ltd, 2003, 737-772.

[3] Montoya Oliver, J.M. El ciervo y el monte: Manejo y conservación (Cervus elaphus l.), *Mundi-Prensa Libros*, Madrid, 1999. (in Spanish).

[4] Baeza, A.; Vallejo, I.; Guillén, J.; Salas, A.; Corbacho, J.A. *J. Environ. Radioactiv.* 2011, 102, 311-315.

[5] ICRP. *Recommendations of the International Commission on Radiological Protection.* Publication 26; Oxford, Pergamon Press, 1977.

[6] Ferenbaugh, J.K.; Fresquez, P.R.; Ebinger, M.H.; Gonzales, G.J.; Jordan, P.A. *Environ. Monit. Assess.* 2002, 74, 243-254.

[7] Åhamn, B. J. Environ. Radioactiv. 1999, 46, 113-120.

[8] Golikov, V.; Logacheva, I.; Bruk, G.; Shutov, V.; Balonov, M.; Strand, P.; Borghuis, S.; Howard, B.; Wright. J. *Environ. Radioactiv.* 2004, 74, 159-169.

[9] Pentreath, R.J. In: *Protection of the Environment from the Effects of Ionizing Radiation*; IAEA-CN–109, IAEA, Vienna, 2005, 411–420.

[10] ICRP. *Annals of the ICRP* 2007, 37(2-4), 133-135.

[11] IAEA. Extent of Environmental Contamination by Naturally Occurring Radioactive Material (NORM) and Technological Options for Mitigation; *IAEA Technical reports series* nº 419, Vienna, 2003.

[12] UNSCEAR, United Nations Scientific Committee on the Effects of Atomic Radiation. *UNSCEAR 2000 Report to the General Assembly, with scientific annexes*; New York, 2000.

[13] Macdonald, C.R.; Elkin, B.T.; Tracy, B.L. *J. Environ. Radioactiv.* 2007, 93, 1-25.

[14] Thomas, P.A.; Gates, T.E. *Environ. Health Pers.* 1999, 107(7), 527-537.

[15] Klevezal, G.A.; Bakhur, A.E.; Kalyakin, V.N.; Krushinskaya, N.L. *Chemosphere* 2001, 42, 61-67.

[16] Macdonald, C.R.; Ewing, L.L.; Elkin, B.T. Sci. Total Environ. 1996, 182, 53-73.

[17] Skuterud, L.; Gwynn, J.O.; Gaare, E.; Steinnes, E.; Hove, K. *J. Environ. Radioactiv.* 2005, 84, 441-456.

[18] Mahon, D.C. *Bull. Environ. Contam. Toxicol.* 1982, 29, 697-703.

[19] Mietelski, J.W.; Gaca, P.; Zagrodzki, P.; Jasińska, M.; Zalewski, M.; Tomczak, M.; Vadja, N.; Dutkiewicz, E.M. J. *Radioanal. Nucl. Chem.* 2001, 247(2), 363-370.

[20] Staaland, H.; Bjørnstad, H.; Pedersen, Ø.; Hove, K. *Rangifer* 1991, 11(1), 17-22.

[21] Landstetter, C.; Wallner, G. *J. Environ. Radioactiv.* 2006, 87, 315-324.

[22] Baeza, A.; Del Río, M.; Jiménez, A.; Miró, C.; Paniagua, J. *Radiochim. Acta* 1995, 68, 135-140.

[23] Farris, G.C.; Whicker, F.W.; Dahl, A.H. In: Lenihan, J.M.A.; Loutit, J.F.; Martin, J.H. (Eds.) *Strontium metabolism*; Academic Press, London, 1967, 93-102.

[24] Gaspar López, E.; Landete Castillejos, T.; Gallego, L.; García, A.J. *Eur. J. Wildlife Res.* 2008, 54, 753-755.

[25] Landete Castillejos, T.; García, A.; Gallego, L. *Bone* 2007, 40, 230-235.

[26] Strandberg, M.; Strandgaard, H. *J. Environ. Radioaciv.* 1995, 27, 65-74.

[27] Tiller, B.L.; Poston, T.M. *J. Environ. Radioactiv.* 2000, 47, 29-44.

[28] Schönhofer, F.; Tataruch, F.; Friedrich, M. *Sci. Total Environ.* 1994, 157, 323-332.

[29] Pokorny, B.; Ribaric-Lasnik. C. *Envirom. Poll.* 2002, 117, 35-46.

[30] Åhman, B. *Rangifer* 1994, 14(1), 23-28.

[31] Steinnes, E.; Gaare, E.; Engen, S. *Sci. Total Environ.* 2009, 407, 3905-3908.

[32] Tracy, B.L.; Kramer, G.H. *Artic* 2000, 53(1), 42-52.
[33] Johanson, K.J.; Bergström, R. *Environ. P*oll. 1989, 61, 249-1260.
[34] McGee, E.J.; Synnott, H.J.; O'Keefe, C.; Colgan, P.A. *Brit. Vet. J.* 1995, 151, 671-682.
[35] Rissanen, K.; Rahola, T.; Aro, P. Ranfiger 1990, 2 (10), 57-66.
[36] Johanson, K.J.; Bergström, R.; Eriksson, O.; Erixon, A. *J. Environ. Radioactiv.* 1994, 22, 251-267.
[37] Skuterud, L.; Gaare, E.; Eikelman, I.M.; Hove, K.; Steinnes, E. *J. Environ. Radioactiv.* 2005, 83, 231-252.
[38] Zibold, G.; Drissner, J.; Kaminski, S.; Klemt, E.; Miller, R. *J. Environ. Radioactiv.* 2001, 55, 5-27.
[39] Fielitz, U.; Klemt, E.; Strebl, F.; Tataruch, F.; Zibold, G. *J. Environ Radioactiv.* 2009, 100, 241-249.
[40] White, R.G.; Holleman, D.F.; Allaye-Chan, A.C. *Rangifer* 1986, 1, 24-29.
[41] Jones, B.E.V.; Eriksson, D.; Nordkvist, M. *Sci. Total Environ.* 1989, 85, 207-212.
[42] Karlén, G.; Johanson, K.J.; Bergström, R. *J. Environ. Radioactiv.* 1991, 14, 91-103.
[43] Strandberg, M.; Knudsen, H. *J. Environ. Radioactiv.* 1994, 23, 189-203.
[44] Pálsson, S.E.; Egilsson, K.; Pórisson, S.; Magnússon, S.M.; Ólafsdóttir, E.D.; Indridason, K. *J. Environ. Radioactiv.* 1994, 24, 107-125.
[45] Aarkrog, A. *J. Environ. Radioactiv.* 1994, 25, 21-35.
[46] Åhman, B. *J. Environ. Radioactiv.* 2007, 98, 153-165.
[47] Forberg, S.; Odsjö, T.; Olsson, M. *Sci. Total Environ.* 1992, 115, 179-189.
[48] Szerbin, O.; Popov, D.K. *J. Environ. Radioactiv.* 1988, 8, 129-141.
[49] Taylor, H.W.; Svoboda, J.; Henry, G.H.R.; Wein, R.W. *Artic* 1988, 41(4), 293-296.
[50] Zach, R.; Hawkins, J.L.; Mayoh, K.R. *J. Environ. Radioactiv.* 1989, 10, 19-45.
[51] Lowe, V.P.W.; Horrill, A.D. *Environ. Poll.* 1988, 54, 81-87.
[52] Johanson, K.J.; Bergström, R. *Sci. Total Environ.* 1994, 157, 309-316.
[53] McGee, E.J.; Synnott, H.J.; Johanson, K.J.; Fawaris, B.H.; Nielsen, S.P.; Horrill, A.D.; Kennedy, V.H.; Barbayiannis, N.; Veresoglou, D.S.; Dawson, D.E.; Colgan, P.A.; McGarry, A.T. *J. Environ. Radioactiv.* 2000, 48, 59-78.
[54] Åhman, B.; Wright, S.M.; Howard, B.J. *Sci. Total Environ.* 2001, 278, 171-181.

[55] Rissanen, K.; Rahola, T. *Sci. Total Environ*. 1989, 85, 199-206.
[56] Strebl, F.; Gerzabek, M.H.; Karg, V.; Tataruch, F. *Sci. Total Environ.* 1996, 181, 237-247.
[57] Švadlenková, M.; Konečný, J.; Smutný, V. *Environ. Poll.* 1996, 92(2), 173-184.
[58] Baeza, A.; Guillén, J. *Bioremediation, Biodiversity and Bioavailability* 2007, 1, 78-87.
[59] Schell, W.R.; Tobin, M.J. In: Desmet, G.; Nassimbeni, P.; Belli, M. (eds.) *Workshop on the transfer of radionuclides in natural and semi-natural environments*; Elsevier Applied Science, London, 1990, 118-128.
[60] Kiefer, P.; Pröhl, G.; Müller, H.; Lindner, G.; Drissner, J.; Zibold, G. *Sci. Total Environ.* 1996, 192, 49-61.
[61] Ohnuki, T. Radiochim. Acta 1994, 65, 75-80.
[62] Lee, M.H.; Lee, C.W. *J. Environ. Radioactiv.* 2000, 47, 253-262.
[63] Dumat, C.; Stauton, S. *J. Environ. Radioactiv.* 1999, 46, 187-200.
[64] UNSCEAR. United Nations Scientific Committee on the Effects of Atomic Radiation. *UNSCEAR 2008 Report to the General Assembly, with scientific annexes,* New York, 2008.
[65] Voigt, G.; Pröhl, G.; Müller, H.; Bauer, T.; Lindner, J.P.; Probstmeier, G.; Röhrmoser, G. *Sci. Total Environ.* 1989, 85, 329-338.
[66] Staaland, H.; Garmo, T.H.; Hove, K.; Pedersen, Ø. *J. Environ. Radioactiv.* 1995, 29(1), 39-56.
[67] Calmon, P.; Thiry, Y.; Zibold, G.; Rantavaara, A; Fesenko, S. *J. Environ. Radioactiv.* 2009, 100, 757-766.
[68] IAEA. Handbook of parameter values for the prediction of radionuclide transfer in terrestrial and freshwater environments; *Technical Report Series n°* 472, Vienna, 2010.
[69] Hove, K.; Staaland, H.; Pedersen, Ø. *Rangifer* 1991, 11(2), 43-48.
[70] Åhman, B. *J. Environ. Radioactiv.* 1996, 31(1), 29-50.
[71] Staaland, H.; Hove, K.; Pedersen, Ø.; Birke, L.C. *Rangifer* 1998, 18(1), 27-34.
[72] Beresford, N.A.; Barnett, C.L.; Jones, D.G.; Wood, M.D.; Appleton, J.D.; Breward N.; Copplestone, D. *J. Environ. Radioactiv.* 99 (2008) 1430–1439.
[73] ICRP. *ICRP Publication* 108, 2008, 38 (4-6).
[74] Klevezal, G.A.; Sokolov, A.A. Russian *J. Zoology* 1999, (3)1, 224–233.
[75] Kryshev, I.I. *Environmental Risk Analyses for the Ural Radioactive Pattern*. Russian Nuclear Society, Moscow, 1997, 1–210.

[76] Sazykina, T.; Kryshev, I.I. *J. Environ. Radioactiv*. 2006, 88, 11–48. N5.

In: Deer: Habitat, Behavior and Conservation ISBN 978-1-62100-676-3
Editors: A. A. Cahler and J. P. Marsten © 2012 Nova Science Publishers, Inc.

Chapter 3

HUNTED RED DEER MANAGEMENT: A NEW TOOL FOR POPULATION SURVEY

Marc Colyn,[1,2*] Xavier Legendre,[3**] Patrick Barrière,[1,4] Jean-Claude Chardron,[5****] Sabine Bertouille,[6*****] and Pierre Deleporte[7******]***

[1]Université de Rennes 1, CNRS, UMR 6553 Ecobio, Station Biologique, Paimpont, France
[2]Université de Rennes 1, E BGT, Campus de Beaulieu, Rennes Cedex, France
[3]MNHN Paris, Parc de la Haute Touche, Obterre, France
4AICA-CREG, Nessadiou, BP 37 98 870 Bourail, Nouvelle Calédonie
[5]DRAAF, Bretagne pôle forêt-bois, 15 avenue de Cucillé, Rennes cedex 9
[6]Département de l'étude du milieu naturel et agricole - Direction de la Nature et de l'Eau, FSC laboratory, Gembloux, Belgium
[7]Université de Rennes 1, CNRS, UMR 6552 EthoS, Station Biologique, Paimpont, France

* E-mail: marc.colyn@univ-rennes1.fr.
** E-mail: xavierlegendre36@gmail.com.
*** E-mail: tech.creggg@aica.nc.
**** E-mail: jean-claude.chardron@agriculture.
***** E-mail: sabine.bertouille@spw.wallonie.be.
****** E-mail: pierre.deleporte@univ-rennes1.fr.

ABSTRACT

Many wild deer populations in the world are considered as threatened and deserving protection measures for population increase; in contrast, some managed deer populations are presently increasing to the point of becoming a nuisance or a permanent pest. In France, free ranging *Cervus elaphus* populations recently showed a strong increase, the global annual hunting plan rising from 10,000 to 71,000 individuals over the last 30 years. Frequently the local picture was an alternation of population peaks followed by crashes, with an approximate 10-years cycle. Such an evolution results largely from the ineffectiveness of the exhaustive counting methods previously used for deer management. Considering both the importance of estimating vital rates for managing deer populations and the extreme difficulty of obtaining reliable data in the field, we devised a new method to survey populations, in order to provide managers with a simple and effective tool. Our method takes advantage of the rich data sample provided by the annual cull, and the fact that hinds are shot randomly in driven hunts; hence the hind cull is a reliable representation of the hind population age structure. The mandibles of all animals killed by hunters are collected and analyzed, which allows the calculation of the ratio of yearling hinds in the total hind cull. Measuring mandible lengths is also an indicator of the populations physical condition. We conducted a survey of the deer population of Paimpont forest (Brittany, France) over ten years, which showed that the standard management methods underestimated population increases, while our yearling hind ratio allowed to more efficiently diagnose population increase and predict population crash. Our method has also been continuously applied since 2007 for deer management in the Indre District (annual cull of some 2000 deer) with promising results. The yearling hind ratio in culls is a simple management tool, dispensing with difficult field counts and allowing for gradual annual correction of management plans. Our study also demonstrated that this approach is applicable in this country on the scale of a whole district, given ad hoc imposed legislation and active contribution of hunting associations and management authorities.

Keywords: red deer, population management, cull control, *postmortem* analysis of hinds, age structure, index of physical condition

INTRODUCTION

Cervids are geographically widespread and represented by 19 genera and 51 species (Wilson and Reeder 2005, Ludt *et al.* 2004). The genus *Cervus* comprises 2 species (*C. elaphus* and *C. nippon*) distributed in North America, Europe, Asia and in a limited part of Northern Africa (Geist 1999). *C. elaphus* has been introduced in Australia, Argentina and New Zealand (the latter populations even becoming a permanent pest) while *C. nippon* was notably introduced in Europe, where it unfortunately hybridizes with *C. elaphus* in different countries (Abernethy 1994, Goodman *et al.* 1999, Senn and Pemberton 2009).

The red deer (*Cervus elaphus*), comprising four subspecies, is widespread in Eurasia and extends southward to Northern Africa (Ludt *et al.* 2004). The subspecies *C. e. barbarus* is the only threatened one (in Morocco, Tunisia and Corsica). In Europe, this species is well represented and in many countries populations are expanding as a consequence of changes in deer management practices during the past decades.

Populations of *C. e. elaphus* had been exterminated in most of the French regions when, during the second half of the twentieth century, deer from Chambord National Reserve were reintroduced in numerous forested areas. In 1979 the application of a Deer Management Plan throughout continental France has favored a progressive increase in deer populations so that they are now on the increase in many regions (ONCFS 2010). Presently, the Indre District shows the highest annual deer cull per forested area surface (15 per 1,000ha). Over the whole of Continental France, the annual Deer Management Plan (DMP) increased from some 10,000 individuals in 1979 to 71,000 in 2009, *i.e.* the annual DMP increased sevenfold in 30 years (Figure 1). This evolution obviously reflects a sharp global population increase over this period.

Red deer are now present in 83 of the 92 continental French districts, where they expanded their range from 7.2 million ha in 1985 to 13.7 million ha in 2005, notably in mountainous regions where they have reached high densities at altitudes of over 1,500 m (Adrados 2002, Pfaff and Sainte-Andrieux 2008, ONCFS 2010). In many regions, the abandonment of agricultural land favored the development of waste lands allowing a general expansion of deer populations into new areas outside their traditional forest ranges.

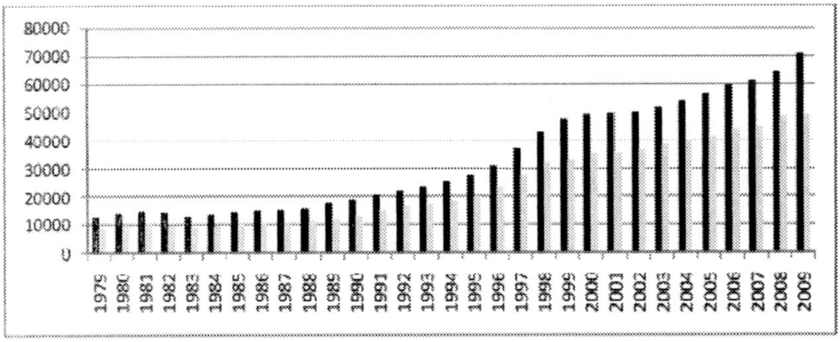

(source: ONCFS).

Figure 1. Annual free ranging deer hunting plan and cull in France from 1979 to 2009. Black blocks: planned cull; grey blocks: declarative cull.

Increasing deer populations is the general trend in Europe; France, with an annual cull of 3.2 red deer per 1,000 ha of forested area, stands only at the sixth rank behind Great Britain (23 red deer per 1,000 ha), Denmark, Germany, Belgium and Luxemburg (6 per 1,000 ha). However, in some French areas, the current increase in local deer populations raises management problems associated with deer damage to farm crops and tree plantations. Even if deer density in France is lower than in neighboring countries, farmers and forest managers are not accustomed to such population levels; fearing serious damages, they are demanding an efficient DMP for strict deer control.

The problem is double-sided, depending on the foresters or the hunters points of view. On the one hand, French forests are frequently composed of single species plantations of fast growing coniferous tree species of exotic origin, which are not favorable in supporting high deer densities. Some forest managers suggest that current red deer population levels would threaten biodiversity, even in deciduous forests. At least it must be admitted that an uncontrolled deer population increase would most likely become a problem for this type of forest management: by limiting forest regeneration, damaging young plantations and finally reducing commercial timber production. Damage to agricultural crops is another problem: deer are strongly resented by farmers but while they are financially compensated by hunting associations, foresters are not. It appears that hunting managers are facing great difficulties in their efforts to stabilize population numbers: deer management tools which have been used for more than thirty years have not prevented recurrent cycles of demographic explosion followed by population crashes. It also seems that some hunting managers are mostly preoccupied by a high annual cull, whatever the consequences to crops and trees.

In this chapter, we present the following: an assessment of deer management practice in France during recent decades, the new method and protocols, and two case studies applying this approach on free ranging hunted populations.

Deer Management in France During the Recent Decades: An Assessment

Since the generalized application of the DMP in 1979, recommended management rules for free ranging deer have been resting on two unshakable principles: 1) attempting to plan an annual cull of 25 % of animals present on the territory after births, which supposedly compensates for the annual population increase; and 2) population monitoring using a series of indirect indicators.

The first rule is not applicable for it is now well known that red deer populations cannot be counted in forested areas, even approximately. Most management problems precisely came from this impossibility.

The second rule, which is more or less efficiently applied depending on the goodwill and funding capacities of local DMP managers, consists of monitoring of populations using principally two bio-indicators: the kilometric index of abundance KIA (from counting by night using spotlights) and the flora pressure index FPI. These indicators provide useful alert concerning the impact of deer on the environment, but unfortunately this frequently comes too late after the start of the concerned population fluctuations. These estimations are completed by sight counting (using a combination of silent drives and static observers), counting deer heard belling at rutting stands during the rut and farm crop damage index (FDI). Unfortunately, these two indicators are difficult to use accurately: counts require great care in protocols to insure repeatability and comparability, while crop damage may be dependent on local annual forest food resource limitations independent of real population fluctuations. Moreover, the imprecision of these indicators is amplified by the species' gregarious way of life, large home ranges and ability to travel long distances. Also, they provide no information on regional population structures and dynamics, and particularly not on hinds that are the principal reproducers, no more than on the quantity and quality of annual cull.

For example, the study we ran on the Broceliande forest area in Brittany (9,000 ha) for 12 years with the Deer Taskforce of Ille et Vilaine District demonstrated that the application of the standard protocols using the series of biological indicators recommended by national institutions did not allow for satisfying population management; they did not allow the prediction of the obvious, strong population increase (Figure 2).

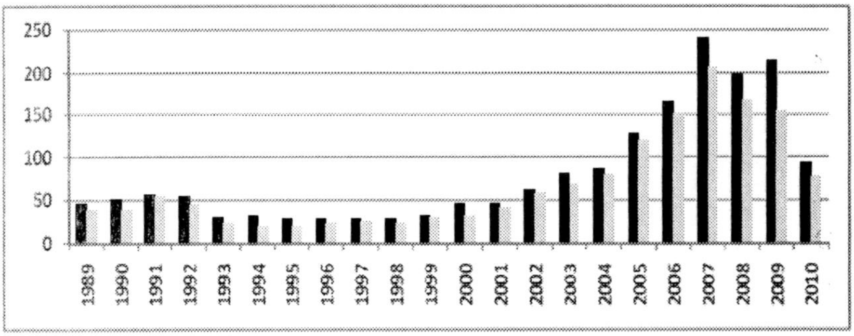

(source: Deer Task Force 35).

Figure 2. Annual deer hunting plan and cull on Paimpont unit (Brittany) from 1989 to 2010. Black blocks: planned cull; grey blocks: cull declared by hunters.

It appeared that when indicators provided a significant signal of population increase it was already too late to regulate the population without taking drastic emergency actions to reduce the population.

Currently many experts are deploring a serious lack of sensibility of methods for populations monitoring and management (Bertouille and de Crombrugghe 1995, 2002; Bertouille *et al.* 2003; Azorit *et al.* 2003; de Crombrugghe 2003). More than this, the collection of cull numbers is based on declaration statements, which are unverifiable. Standard indicators used in establishing DMPs are showing their limits and did not allow for sustained maintenance of fairly stable populations inside limits which are tolerable for forestry management (Bonenfant *et al.* 2005).

As a result of this lack of reliable population statistics, actual management practice only consists of case-by-case action based on very approximate data and forecasts; notably, declared cull data from hunters which are non verifiable. The classic scenario is as follows: 1) there are indications that a population is growing; 2) action is taken to implement a population monitoring plan with environmental damage indicators which will confirm population increase; 3) finally, in order to satisfy all concerned parties, shooting quotas are simply raised; which does not necessarily give the desired immediate results because there is neither control of the effective cull nor the analysis of its impact on the population. Often there is a strong delay between events, diagnosis and decisions, so that regulatory actions are not necessarily appropriate for the new population dynamics and new environmental conditions. As a result, population crashes follow phases of uncontrolled population explosions (Figure 3) and some ten years later a new similar cycle produces similar effects. Also, such a chaotic management has a deleterious impact on population structure that remains imbalanced in terms of proportions of age classes. Such effects are unfortunately well documented on a national scale, as revealed by the jagged curves of annual local DMP numbers (ONF 1999). As a recent illustration, official statistics from ONCFS for the last 2009-10 hunting season show an increase in cull numbers of 52% to 63% in three districts, while there is a decrease in 23% to 38% in five other districts, such highly unstable hunting cull figures indicating lasting management difficulties in numerous territorial units.

Figure 3. A very large herd of free ranging red deer hinds, illustrating a local uncontrolled population explosion (J. P. Villermé, 2006).

In summary, it clearly appears that the DMP developed since 1979 favored a generalized, progressive restoration of populations, which demonstrates that we are able to restore a population shortage; conversely, it seems that we are currently unable to manage population abundance: the current recommended management methods do not allow for long term population stability at the desired level by local managers and hunters in charge of the problem.

These observations reveal a lack of reliable tools for assessing deer population size. Moreover, hunting practices are traditionally focused on males, while reproducing hinds that are major factors of population dynamics are generally ignored in population management, particularly in France (Bonenfant *et al.* 2003, 2005; Bonnet, 2005). In other European countries, several authors have studied deer population structure and dynamics (Langvatn and Loison 1999; Albon *et al.* 2000). However, methods for population monitoring in the field are complex and very expensive (Bonenfant 2002, Bonenfant *et al.* 2005); hence they can hardly be implemented for current use on a large scale.

Consequently: 1) management is mainly done via a monitoring of males, and annual exhibitions of antler trophies are a frequent event with hunters associations; but such surveys typically show recurrent phases of fewer of old males, hence revealing the failure to produce a sustainable number of full grown and old stags, which is precisely one of the main goals of deer hunting management; 2) in some areas, the almost total lack of monitoring of hunting pressure on hinds and fawns led to a spectacular demographic explosion of herds or to their splitting and spreading when suitable space was available; 3) on a national scale, there are no reliable data on annual hunting statistics, available data being as a rule issued from uncontrolled declaration reports without reliable distinction between the fawn, hind and stag classes (ONCFS 2010).

Based on our experience (see case studies below), we suggest to develop a new method, easily used in the field, allowing for reliable monitoring of deer populations, so that managers can maintain red deer densities at the desired levels. Contrary to many current claims that red deer are too numerous in this country, we think that it is possible to maintain present populations, and even in some places to raise their numbers, provided that efficient control would be implemented, and this even without knowledge of exact population numbers. The observation of loss of physical condition in a deer population may justify decreasing its numbers, because this is the symptom of a developing imbalance between the local population and its environment. Of course this approach would require the implementation of a monitoring of physico-physiological parameters like skeletal measures, weight or state of pregnancy of young hinds.

We want to provide wildlife managers with a reliable method, specifically applicable to every territory, allowing them to take well informed decisions and to implement graded actions by all partners involved (be it for stabilizing or modifying population size); hence preventing recourse to drastic, emergency corrective measures that are detrimental to population structure and hence to regular hunting practice.

THE 'YHR' METHOD

We present here what we call the "YHR" method, becaused it is mainly based on the measure of the Yearling Hind Ratio in annual culls (this method has occasionally been labeled "Colyn's method" in some French circles of deer population management). We oriented our studies toward the production of a new management method applicable to most of situations encountered in French deer ranges. The method and corresponding technical protocols must accommodate the extreme diversity of habitats (lowland or mountain forests, rich or poor soils…) and to the diversity of behavior of local populations; also, they must fit to the complexity and fragmentation of hunting blocks. We cannot use of density estimations, because we have no access to realistic numbers, nor the state of pregnancy of hinds, which fluctuates with time for a given population. Of course it is possible to examine closely hind genitalia, particularly in vast hunting territories under control of a unique managing authority. But this requires special capacities from the analyst, and thus this protocol cannot be generalised to investigations by all hunters at the scale of a district; *e.g.* the Indre District alone has to deal with more than 740 files of owners of deer hunting rights, and of course some of them will not have the required skills.

The management tool that we developed at University of Rennes 1, in collaboration with MNHN Paris and the Wild Fauna and Cynegetics Laboratory of Gembloux (Belgium), focuses on 3 points: 1) exact census of cull; 2) computing yearling hind ratio (YHR) in annual hind cull; 3) measuring a physical condition index. Managing must be performed as much as possible on coherent population units.

1) Collect an exact census of all individuals culled annually (fawns, hinds and stags): this is performed through the collect and observation of their mandibles, and this process is mandatory to hunters and imposed by the district legislation;

2) Monitoring the yearling hind ratio in cull (YHR): this is the ratio of the number of yearling hinds to the total number of hinds taken. The YHR is a biological parameter supposed to be linked to population structure and dynamics: from year to year, and all things being equal otherwise, an increase in YHR means that a greater proportion of fawns than previously have been

allowed to become yearling hinds; hence the number of fawns taken was lower than that which would have given a stable YHR. The YHR can be calculated separately for every independent hunting area and it gives useful information provided that culls are: i) the fawn/hind ratio in cull (empirically fixed as approximately equal between fawns and hinds) is fairly stable from year to year (hence allowing direct comparison of annual statistics); ii) random for hinds, *i.e.* without particular shooting recommendation for this category, so that hinds of all age classes are shot in proportion of their share in the population, and the cull is thence representative of the age class structure for hinds; for fawns, hunters generally do not discriminate between males and females, hence the cull is roughly balanced between the sexes in this category (it should be noted that driven hunting is by far the most widespread deer hunting method in France, and this way of shooting hardly allows discrimination in sex categories among fawns and age categories among hinds).

The YHR, directly controlled by biologists, allows, year after year, to assess the impact of the cull on the whole population. Yearling hinds are surviving fawns from the previous year; hence, the trend (rising or decreasing) of the YHR on a given year provides retrospective information, one year later, on the effect of the actual cull of the previous year on the annual evolution of the population size. A population under adequate stabilizing hunting pressure would show a stable age structure among randomly taken hinds, hence a stable YHR. If all fawns of year *t* are taken by hunters, no yearling hinds are available for hunting on next year *t*+1. Conversely, if no fawn is taken on year *t*, the proportion of yearling hinds will show an increase in the hind cull statistics on year *t*+1. Of course, real situations vary between these two unlikely extremes. Supposing that the expectation is a stabilization of the population size, observed stability of the YHR indicates that the previous year cull of fawns and hinds was correct; a significant increase in YHR means that the population tends to grow, and conversely. Without even analyzing the pregnancy status of yearling hinds, this method allows for diagnosing the real impact of annual culls on the evolution of population structure, hence on population dynamics (while the exact population size or the number or births or the specific mortality from different causes will always remain unknown).

Bonenfant *et al.* (2005) argued that the young female ratio cannot be used for reliable prediction of ungulate population dynamics, notably because the pregnancy status of hinds can undergo strong and unpredictable fluctuations. Our approach makes use of YHR in a different way. We are not aiming at precise prediction based on estimated parameters; we are performing precise retrospective analysis based on reliable parameters taken from post-mortem data of a highly representative sample. In other words, we are not trying to adjust the annual DMP according to estimations of its future impact on the predicted population dynamics; we are correcting the DMP according to the observed effect of the real impact of previous annual DMPs. The correction occurs one year after the implementation of the DMP, and this delay seems sufficiently short to allow efficient planning where hunting is the main factor influencing population dynamics, as will be seen from our experimental application.

3) The survey of an index of physical condition: the length of the left mandible, which is correlated with individual weight in young animals (Bertouille and de Crombrugghe 1995, 2002; Bertouille *et al.* 2003; Azorit *et al.* 2003; de Crombrugghe 2003). All things being approximately equal otherwise in terms of environmental parameters, this index allows an assessment of the evolution of population physical condition, particularly for the young individuals, because individual weight is notably positively correlated with food quantitative availability and quality, hence with population density (at least for high population densities).

This method is designed for managing population units of a sufficient size, as much as possible: fluctuations of a too small population are difficult to predict, while tiny territories below the size of the regular home range of deer should obviously not be considered as management units. Also, the population should be relatively insulated from neighboring populations; in case of a high rate of exchanges with other populations, it is recommended to enlarge the geographical extension of the studied and managed area.

PROTOCOL

1) All the mandibles of the cull are collected from the hunters once a month by agents of the district hunting managers. They are counted and sorted by classes (fawns, hinds and stags) and checked against the unused tags brought back by hunters so that no error is possible concerning the actual annual cull. Unlike declaration statements-based approaches, this is the only way to know with certainty the exact cull in all deer categories.
2) Yearling hind mandibles are separated from those of older hinds, using as criteria the presence in yearlings of a tri-lobulate lacteal premolar p_3 and the presence of only the M_1 and M_2 molars. This allows

the calculation of the YHR for the current year, to be compared with those of the preceding years for estimating trends in global population dynamics.

3) These mandibles are then measured between the posterior limit of the mental foramen (*foramen mentale*) and the posterior mandible angle. This allows the calculation of the mean mandible length for the current year, to be compared with those of preceding years for monitoring the evolution of physical condition.

The annual cull can be considered as a sampling effort sufficient for the main DMPs. The annual cull under fairly stable population and environment conditions is supposed to be around 25% of the target population (ANCGG 2004). Even if, extraordinaryly, the actual cull would happen to be only some 15 % of this population, such a sample would nevertheless be highly representative, and quite reliable for management purposes.

TWO CASE STUDIES

In 1998 we started a study of the DMP applied on red deer in the Broceliande Forest in Ille et Vilaine District, in Brittany Region (Colyn and Chardron 2005). For some 9 years, the use of the standard series of indicators provided the usual unsatisfying results (Figure 2); hence we began to work on the development of new methods from 2005 onwards (Colyn 2008a, 2008b). Besides this, the new method was experimentally applied on forest areas in the Indre District (Center Region) which is one of the French districts with the highest annual deer cull. This essay gave convincing results, which led the district administration to impose its application on the whole range of the Indre District (through Arrêté Préfectoral, from 2007 onwards).

Study on Broceliande Forest (Brittany)

Broceliande Forest comprises 2 distinct game management units. One is the private and public areas part, covering nearly 9,000 ha (hereafter called Paimpont); the other is a military area, covering 5,000 ha (hereafter: Coëtquidan).

We selected Brocéliande Forest to test the practicality of our protocols under real hunting management conditions, because of particular qualities: the local deer population is fairly genetically isolated from neighboring ones (Dellicour *et al.* 2011), and it is of moderate size so that we expected: 1) fast reactions of the population to changes in DMP, and 2) easy and accurate checking of annual cull. Since 1998 we developed an annual survey of standard DMP on the whole Brocéliande forest range. Exhaustive annual collecting of mandibles from hunters allowed an accurate numbering of the cull and corresponding analysis.

In 1998, deer numbers were very low following a crash in the population on the Paimpont unit (Figure 2). Using the classic indicator survey method, managers remedied the population shortage quite successfully, but were unable to predict the sharp demographic expansion observed as soon as in 2004, and this despite the progressive increase, from year 2000 to 2004, of shooting quotas (45 to 85) and of the actual cull (32 to 79). During the 3 following years, obviously the local DMP kept on running after an explosive population increase, passing from 129 to 240 (quotas) and from 119 to 206 (controlled actual culls). Then the district administration, anxious to safeguard the interests of farmers and forest managers, imposed extreme regulation measures. Deer populations have obviously fallen down quickly to their lowest from 2008 to 2010 (Figure 2). This local pattern of chaotic DMP on Paimpont reflects the general tendency in most of French deer hunting territories. Contrastingly, managers of the Coëtquidan unit did not follow the sharp increase in shooting quota, and the local deer population seems to have remained fairly stable according to standard indicators.

We have applied our YHR method on the local deer population over the recent years. On the Paimpont unit, the YHR in the annual cull was 22 % in 2007-08 and 14.5 in 2008-09 (Table 1). This reduction is the result of the important cull during the hunting season 2007-08. Culls were exceptionally high while this was the first time that we observed such a fall in YHR. Interestingly, our management tool shows inverse results on the Coëtquidan management unit, in the same forest area and for the same period, hence under the same environmental and climatic conditions (Table 1). This is explained by the management politics on this unit, opposed to that of the Paimpont unit, and applying moderate quotas, because the managers wished to increase local deer numbers. We can observe that this management strategy entailed a high YHR for several years on this unit.

Table 1. YHR, hind cull and fawn cull in two successive hunting seasons in Broceliande region (Brittany). The neat drop in YHR on Paimpont in 2008-09 clearly shows the impact of the important cull in 2007-08 while the neighboring Coëtquidan unit shows stable, moderate fawn cull and increasing YHR

	Paimpont		Coëtquidan	
	2007-08	2008-09	2007-08	2008-09
YHR	22%	14.5%	26.1%	35%
Hind cull (Nb)	81	59	22	20
Fawn cull (Nb)	72	55	21	20

As a conclusion, according to our method, the sharp decrease in YHR *e.g.* between hunting seasons 2007-08 and 2008-9 shows that the population underwent a strong decrease in total number, due to high fawn and hind culls. Conversely, the Coëtquidan cull was compatible with a slight population increase, as reflected by the sensible rise in YHR over the same period.

Study on Indre District (Centre Région)

The Indre District Administration had become fully conscious of the fact that red deer populations inexorably increased year after year despite a local increase in shooting quotas of 500% in 20 years (from 728 animals in 1990 to 3791 in 2010). Herds were conspicuously becoming larger and larger and the numbers of antler trophies in annual exhibitions increased regularly. Hence, from the 2007-08 season on, the Administration decided to impose the collecting of mandibles throughout the district territories, so that the YHR method could be applied. During the last 4 years, we analyzed 7812 mandibles issued from 5 independent game management units (Gâtinais; Mitoyen Indre et Loire; Brenne; Petite Brenne; Est et Champagne) defined according to spatial distribution of connected forest blocks and to what is known of circulation habits of the animals moving from one block to another.

Control of The Actual Cull

At the end of the first experimental season (2007-08), the analysis of mandibles of fawn and hinds revealed a discrepancy of some 20% between the controlled cull score of the current year and the declared reports of the previous years. The current actual cull was neatly below those declared previously, despite similar shooting quotas and hunting practices. Hunters used to regularly take 70% of the allowed quota of stags but only 50% of the fawn and hind quota, while the biased declarations were hiding this important discrepancy, and this has probably been happening for several years. So the decision makers, unaware of this bias, raised approximately the quotas on the basis of vague population data, and could only chase blindly to try and control an ever-growing population. The procedure of exhaustive mandible collection put a remedy to this bias.

Survey of YHR

As an illustration of the method, we show the analysis performed on 1013 mandibles of hinds taken on the Brenne unit for 4 seasons (2007-08 to 2010-11). There is a sensible decrease in the YHR from 26% to 22.1%. It is clear that the hunting pressure on fawns and hinds was exceeding what would have allowed a stable population density. Similar trends were recorded for the other 4 units in the District, which indicates that the DMP allowed to efficiently and progressively reduce these populations over several years. These results fitted quite well with the stated goals of both the District Federation of Hunters and District Territories Administration, which wanted to improve management practices.

Survey of Physical Condition

Up to this date, our annual monitoring of the mandible index of physical condition allowed the diagnosis of two unpredictable features: 1) a very neat and stable discrepancy between the mean yearling hind size of different populations, and 2) inter-annual differences concerning several populations.

1) Inter-population variations: yearling hinds from the 5 management units of the district show contrasted mandible length statistics (Table 2). Larger animals came from the three units with intensive farming practices (Gâtinais, Mitoyen Indre et Loire, Est et Champagne), while those of smaller size came from units of poor forest habitat and crop productions (the two Brenne swampy areas). More precisely, Brenne differs significantly from Est et Champagne, and petite Brenne differs significantly from the 3 non-Brenne

units (ANOVA, Bonferroni comparisons, p<0.05). These data show that physical condition is correlated with local habitat quality, hence the requirement of treating separately every single established population.

Table 2. Compared yearling hind mandible lengths from hunting season 2007-2008 to 2010-2011 in five hunting units. The exceptional raise in mandible length in 2009-2010 yearling hind cull in three units is likely linked to the exceptionally high oak acorn production in fall 2007 (adding to the effect of decreasing population densities)

	Gatinais	Mitoyen Indre et Loire	Brenne	Petite Brenne	Est et Champagne
2007-2008	209.40 ± 8.29	209.29 ± 8.46	202.44 ± 9.12	206.16 ± 7.56	209.25 ± 6.74
2008-2009	209.71 ± 4.57	205.56 ± 4.16	205.35 ± 8.46	205.20 ± 8.52	208.30 ± 8.18
2009-2010	216.83 ± 8.33	215 ± 7.13	208.18 ± 8.17	201.71 ± 9.69	211.66 ± 6.26
2010-2011	208.29 ± 10.98	212 ± 6.95	205.34 ± 8.35	205.52 ± 7.75	211.21 ± 9.32

2) Inter-annual variations: the lengths of yearling hind mandibles taken in 2009-10 were significantly larger than those of 2008-09 (ANOVA, Bonferroni comparisons, p<0.05), and exceeded those from the 2 other years culls (Table 2), and this inside all units with the exception of Petite Brenne. This increase in size, hence improving in physical condition, likely results from the exceptional production of oak acorns during the autumn of 2007 (Vilmorin information service, pers. comm.), *i.e.* the period when these yearlings were already at the fetus stage, while the Petite Brenne forest unit has a relatively poor oak density and the local deer population was probably facing strong food competition for acorns from the relatively high wild boar density. Such observations are important for game management, because ovulation in yearling hinds is triggered by body weight (hence body size); hence this big sized female cohort detected in 2009-10 is very likely to contribute significantly in increasing the population from spring 2010 on. This constitutes an element of information for DMP adjustment: an exceptionally good reproductive season should not be confused with the longer term trends of population size.

Otherwise, no other significant inter-annual differences in mandible lengths were observed during these four years on these five units, despite the regular decline in deer populations indicated by the evolution of YHR. This is suggesting that these population fluctuations resulting from DMPs were sufficiently gradual that populations never left (nor reached) a state of strong disequilibrium with the available environmental resources.

CONCLUSION

The new management protocol was imposed by the Indre District Administration for the last 4 years, thus allowing a controlled assessment of the effective cull. The great simplicity of the protocol allowed the efficient participation of local partners. Notably, nearly all the partners involved and organized through the District Hunting Federation forwarded the mandibles, declaration forms and unused tags by the due date. For this 4 year period, nearly 95% of the data were available just a few days after the closure of the hunting season, which clearly demonstrates that this protocol was not an obstacle for the application of this method, be it for the strict control of the actual cull or for *postmortem* measurements on a very large sample. This testing of the method in a whole District also showed that it is quite possible to analyze around 2,000 mandibles a year quite quickly, so that the final audited cull data are available at the beginning of April, *i.e.* at the time when managers are starting to discuss next year quotas. This timing is of course a crucial requirement for efficient management.

After only 4 years of implementation, the YHR method for DMP allowed accurate predictive population management based on biological parameters representative of the population structure, using the evolution of YHR as an indicator of population dynamics. Our surveys are currently ongoing so that the method can be further empirically assessed in the long term.

Classical indicators of ecological impact developed by the monitoring organization *Observatoire de la Grande Faune et de ses Habitats* can play their role as alert signals of an important imbalance between population densities and environmental resources, but such tools are unsuitable for early and finely tuned diagnostic of fluctuation trends in population size.

Bonenfant *et al.* (2002, 2005) underlined both (i) the importance of knowing vital statistics (particularly reproduction, rate of gravid yearling hinds, and survival) for managing exploited populations of ungulates, and (ii) the difficulty of obtaining reliable population counts (particularly if using capture-mark-recapture and long term monitoring of marked animals). We understand that field counts on free ranging animals are very difficult, costly and largely unreliable; this is precisely the reason why, for managing hunted populations, our method takes advantage of the excellent population sampling provided by hunters by random shooting of hinds in driven hunts. The annual cull allows for accurately measuring the YHR, and this YHR of year *t* integrates both the impact of birth rate and fawn survival of year *t*-1, all relevant factors having already had their influence, including of course the major mortality factor due to the hunting cull itself. Monitoring the YHR in the culls is thus a simple, integrated and reliable management tool, not requiring unreliable field counts and counting gravid yearlings. It allows the gradual annual correction of Deer management plans that in France are the major factor of hunted deer population size fluctuations. Our study has also demonstrated that this approach is perfectly applicable in this country on the scale of a whole district, given the ad hoc imposed legislation and active contribution of hunter associations and managing authorities.

ACKNOWLEDGMENTS

We thank the Ille et Vilaine District Deer Taskforce (GTC 35), the Indre District Federation of Hunters (FDC 36), ADCGG 35 and ADCGG 36 for technical and organizational support, and hunters from Paimpont and Indre District for useful contribution to the programs. We are particularly grateful to Véronique BIQUAND for statistical analyses; to Guy BERGUE and Guy BONNET (ANCGG) and Jean-Paul GROSSIN for fruitful discussions and indefectible support; to J. P. VILLERME for allowing to use his hind picture; and to Guillaume BRUNETEAU (Domaine National de Chambord) and Vincent BRIGODE (Fordie Estate, Scotland) for useful comments on a first draft of this chapter.

REFERENCES

Abernethy, K. (1994). The establishment of a hybrid zone between red and sika deer (genus *Cervus*). *Molecular Ecology* (6):551–562.

Adrados C. (2002). Occupation de l'espace et utilisation de l'habitat par le cerf (*Cervus elaphus*) en forêt tempérée de moyenne montagne. Approche au moyen du GPS. *Thèse de doctorat.* Université Paul Sabatier, Toulouse.

Albon S.D., Coulson T.N., Bown D., Guinness F.E., Pemberton J.M. and Clutton-Brock T.H. (2000). Temporal changes in key factors and key age groups influencing the population dynamics of female red deer. *Journal of Animal Ecology* 69: 1099-1110.

ANCGG 2004. Le grand gibier: les espèces, la chasse, la gestion. Paris, Gerfaut.

Azorit C., Analla M. and Munoz-Cobo J. (2003). Variation of mandible size in red deer *Cervus elaphus hispanicus* from southern Spain. *Acta Theriologica* 48(2): 221-228.

Bertouille S. and Crombrugghe de S. (1995). Body mass and lower jaw development of the female red deer as indices of habitat quality in the Ardennes. *Acta Theriologica* 40(2): 145-162.

Bertouille S. and Crombrugghe (de) S. (2002). Fertility of red deer in relation to area, age, body mass and mandible length. *Z. Jagdwiss* 48 (suppl.): 87-98.

Bertouille S., Buchet R. and Villers M. (2003). Les constats de tir et les relevés d'indices de condition physique comme outils de gestion. *Forêt Wallone* 3: 5-11.

Bonenfant C. (2008). Dynamique des populations des grands herbivores : application au cerf élaphe (*Cevus elaphus*). In: Acte du colloque, Symposium Cerf Dijon: 85–96.

Bonenfant C., Gaillard J.-M., Klein F. and Loison A. (2002). Sex- and age-dependent effects of population density on life history traits of red deer *Cevus elaphus* in a temperate forest. *Ecography* 25: 446–458.

Bonenfant C., Gaillard J.M., Loison A. and Klein F. (2003). Sex-ratio variation and reproductive costs in relation to density in a forest-dwelling population of red deer (*Cervus elaphus*). *Behavioral Ecology* 14(6): 862-869.

Bonenfant C., Gaillard J.-M., Klein F. and Hamann J.-L. (2005). Can we use the young female ratio to infer ungulate population dynamics? An empirical test using red deer Cervus elaphus as a model. *Journal of Applied Ecology* 42: 361–370.

Bonnet G. (2005). La biche, on la gère aussi. *Connaissance de la Chasse* 346: 54-63.

Colyn M. and Chardron J-C. (2005). La gestion du cerf élaphe sur le massif de Brocéliande (Bretagne) : bilan de sept années de gestion (1998 – 2004). *Grande Faune* 105: 9-18.

Colyn M. (2008a). La gestion du cerf: les indicateurs dans la balance. *Connaissance de la chasse* 66: 66-70.

Colyn M. (2008b). L'examen biologique des réalisations, un outil incontournable pour la gestion de l'espèce. In: Acte du colloque, Symposium Cerf Dijon: 97.

Crombrugghe (de) S. (2003). Outils de gestion des populations de cerfs et de chevreuils à l'aube du XXIème siècle. *Forêt Wallone* 3: 3-4.

Dellicour S., Frantz A. C., Colyn M., Chaumont F., Bertouille S. and Flamand M.C. (2011). Population structure and genetic diversity of red deer (*Cervus elaphus*) in forest fragments in north-western France. *Conservation Genetics.* In press.

Geist V. (1999). Deer of the World: their Evolution, Behaviour and Ecology. Swan Hill Press, UK.

Goodman S. J., Barton, N. H., Swanson, G., Abernethy, K. and J. M Pemberton. (1999). Introgression through rare hybridization: A genetic study of a hybrid zone between red and sika deer (genus Cervus) in Argyll, Scotland. *Genetics* 152(1): 355–371.

Langvatn R. and Loison A. (1999). Consequences of harvesting on age structure, sex ratio and population dynamics of red deer *Cervus elaphus* in central Norway. *Wildlife Biology* 5 (4): 213-223.

Ludt C.J., Schroeder W., Rottmann O. and R. Kuehn. (2004). Mitochondrial DNA phylogeography of red deer (*Cervus elaphus*). *Mol. Phyl. Evol.* 31: 1064-1083.

ONCFS (2010). Tableaux de Chasse ongulés sauvages –, *Supplément Faune Sauvage* 289.

ONF (1999). Evolution de la densité du gibier de 1980 à 1994 : essais de synthèse. RENECOFOR 319p.

Pfaff Emmanuelle and C. Sainte-Andrieux. (2008). Développement récent de l'espèce en France. In: Acte du colloque, Symposium Cerf Dijon: 10-18.

Senn, H.V., Pemberton, J.M., (2009). Variable extent of hybridization between invasive sika (Cervus nippon) and native red deer (*C. elaphus*) in a small geographical area. *Molecular Ecology*, on-line early. In press.

Wilson and Reeder. (2005). *Mammal species of the world: a taxonomic and geographic reference.* Smithsonian Institution Press.

In: Deer: Habitat, Behavior and Conservation ISBN 978-1-62100-676-3
Editors: A. A. Cahler and J. P. Marsten © 2012 Nova Science Publishers, Inc.

Chapter 4

LEAD POISONING FROM AMMUNITION: SEARCHING FOR THE MAGIC BULLET

Jeff Knott
Royal Society for the Protection of Birds

ABSTRACT

Bullets used for deer shooting have traditionally and in many areas continue to, contain lead. Lead is a known toxin and poisoning can occur via several routes, including ingestion. Lead poisoning of waterfowl has led to regulations and agreements limiting the use of lead shot over wetland areas in many countries. More recent evidence has shown further significant issues in terrestrial habitats and affecting a wider range of species, notably birds of prey and other scavengers via secondary poisoning. These problems also extend to fragments of lead bullets. The standard butchering process often does not remove these ammunition fragments (from both shot and bullets) and contamination of the human food chain resulting from this can be sufficient to lead to increased risk of some health problems, particularly in high risk groups. Efficacious alternatives to lead bullets and shot have been produced and are readily available in most areas. Both voluntary and regulatory approaches have been attempted to reduce exposure to lead from ammunition sources, but these have so far been inadequate with reference to terrestrial areas and lead bullet use. Voluntary and regulatory approaches each have their costs and benefits and the most appropriate course of action will likely vary between situations and locations. While it is difficult to highlight the most appropriate response as a result, there is little doubt that further response is required by stakeholders, including Governments, if the problem is to be effectively addressed.

Lead has long been commonly used in ammunition and continues to be the base substance for most bullets used in a variety of situations worldwide. Its high density and relatively cheap price lent itself to production of accurate ammunition types able to transmit sufficient power to achieve rapid death of the target. For this reason, lead has traditionally been the most widely used base substance in manufacture of most ammunition types, including shot gun pellets, air gun pellets and rifle bullets.

Despite this wide use, lead is also known as a significant toxin to many living organisms. A wide range of toxic effects of lead are known. Injurious effects of lead are numerous and can be caused by sustained low levels of exposure. For example, in humans, there is evidence of permanent adverse effects of lead on cognitive function in children with blood lead levels below 10 μg dL^{-1} (Canfield *et al.* 2003; Lanphear *et al.* 2005; Chandramouli *et al.* 2009). A tolerable weekly intake of 25 μg/kg b.w. is currently recognised, although the European Food Safety Authority's Panel on Contaminant in the Food Chain has concluded that this is no longer appropriate, and declines to suggest a new tolerable weekly intake level, as there is no evidence for a threshold for critical lead-induced effects (EFSA, 2010).

Several routes of exposure have been identified for humans. Gaseous lead compounds, for example as produced through combustion of lead petrol, can enter through the respiratory system, while solid compounds can be ingested and accumulate in body through the digestive system.

Lead Shot over Wetlands

As a result of widespread use in wildfowling activities across much of the globe, significant amounts of spent ammunition were deposited in wetland ecosystems. Poisoning from lead shot has been well documented as a cause of mortality in waterbirds across the globe (eg Pain 1992). Densities of spent shot in the upper substrates can be extremely high, with up to 30 shot/m^2 being reported on some UK flight ponds (Thomas, 1982) and densities of over 100 shot/m^2 reported from some wetland areas in Denmark (Peterson and Meltofte, 1979), France (Pain, 1991) and Spain (Mateo et al, 1998; Bonet et al, 1995; Mateo et al, 2007). With ingestion rates of around 6% reported in widespread species such as mallard (Street, 1983), the potential for significant mortality is clear and indeed it has been conservatively estimated that at least 8,000 Mallard (*Anas platyrhynchos*) would have died from this cause between

September and February every year in the UK alone (Mudge, 1983). At least 18 species of waterfowl have shown lead shot ingestion in Europe (Mateo, 2009)

LEAD BULLETS AND TERRESTRIAL HABITATS

Research indicates that fragments from traditional copper-encased lead bullets and lead shot fragments in the carcasses of animals killed by hunters and pest controllers away from wetlands may be a larger hazard to wildlife than was previously thought (Pain *et al.* 2007). Furthermore, birds such as partridges and pheasants ingest spent shot from terrestrial systems (Potts 2005). At least 63 bird species have been documented to have ingested lead or suffered lead poisoning from ammunition sources, including 10 Globally Threatened or Near Threatened Species (Pain *et al.* 2009).

One of the main groups to be affected, principally through accidental ingestion of shot or bullet fragments in unretrieved game, is raptors (Miller *et al.* 2002). For example, in the Utah desert (USA), Bald Eagles (*Haliaeetus leucocephalus*) feeding on jack rabbits (*Lepus* sp.) may frequently ingest shot, with 71% of regurgitated pellets containing one or more pieces of shot (Platt 1976). Golden Eagles (*Aquila chrysaetos*) have been found with lead toxicosis in the European Alps (Kenntner *et al.* 2007) and Spain, where one bird had 40 pellets in its proventriculus (Cerradelo *et al.* 1992). In Germany lead intoxication has been identified as the major cause of death in White-tailed Eagles (*Haliaeetus albicilla*), with 25% of carcasses examined having died because of lead toxicosis; lead from both shot and bullet fragments was implicated (Krone *et al.* 2003). Lead poisoning from ammunition is considered to be the single most important cause of mortality in this population (Krone 2009).

It was previously thought that the majority of bullets stayed in a single mass or as large fragments in a carcass, and could be easily avoided by scavenging birds. However, recent radiographic studies in the USA have shown that small bullet fragments existed up to 15 cm from the main wound channel in tissues of deer shot with lead bullets (Hunt *et al.* 2006). Ninety percent (n=20) of the deer carcasses examined contained bullet fragments which were mostly small (<2mm) and numerous (mean 160 per carcass). This suggests the probability of accidental ingestion of bullet fragments by scavengers is significantly greater than previously thought. *As offal piles are also traditionally left close to the site of the kill by deer hunters in the UK, these findings are clearly relevant to the UK situation.*

As obligate scavengers, vultures are vulnerable to the ingestion of lead shot and bullet fragments. Examples include Eurasian Griffon Vultures (*Gyps fulvus*) poisoned in Spain (Mateo *et al.* 2003) and Turkey Vultures (*Cathartes aura*) in Canada exhibiting bone lead levels indicative of elevated lifetime exposure, most likely as a result of scavenging hunter-shot carcasses (Martin *et al.* 2003). Reintroduced populations of the Critically Endangered California Condor (*Gymnogyps californianus*) suffer markedly from lead toxicosis. Lead poisoning remains the leading cause of death among free-ranging birds, having caused 60% of deaths in the last five years (Watson, 2009). Ammunition residues in rifle- and shotgun-killed animals are the principal source of this contamination (Parish *et al.* 2009, Chesley *et al.* 2009). In the absence of current intensive efforts to treat birds in order to reduce lead-induced mortality, levels of exposure to lead from fragments of spent ammunition are incompatible with the establishment of a self-sustaining condor population in Arizona and Utah (Green *et al.* 2009).

Human Health Impacts

Meat from game animals shot using lead ammunition is a potential source of lead absorption in humans and regular consumption of game meat is associated with increased blood lead levels (Iqbal *et al.* 2009). Exposure of humans to lead occurs by several routes including dietary exposure. Meat from game animals shot using lead ammunition is a potential source. It was previously thought that meat from game shot with lead bullets and lead shot posed a minimal hazard because nearly all of the mass of the projectile remained in one large piece, which either passed through the carcass or was removed during food preparation or at the table.

However, recent radiographic studies in the United States show that edible tissues of deer shot with lead bullets often contained bullet fragments, which were detected up to 15 cm from the main wound channel (Hunt et al. 2006). Ninety percent of the deer carcasses examined contained bullet fragments which were mostly small (<2mm) and numerous (mean 160 per carcass). Many of the lead fragments were sufficiently small and distant from obviously injured tissue that professional butchers did not remove them when trimming venison for human consumption. Eighty percent of 30 deer shot with lead bullets, each of which was processed by a different butcher, gave rise to at least one 0.9 kg package of minced venison containing one or more lead fragments and 34% of packages contained fragments (Hunt et al. 2009). Similar bullet fragment contamination of red deer and wild boar meat has been reported from Poland (Dobrowolska and Melosik, 2008).

Concerns have also been raised regarding lead contamination of the meat of game animals killed using lead ammunition and of gamebirds which had ingested spent lead shot. Breast muscle tissue from some seabirds killed for food using lead shot in Greenland was sufficiently high for sustained consumption to be potentially hazardous to human health, even though all visible shot were removed from the tissues before analysis (Johansen et al. 2004). Presumably this arises because of the presence of small fragments which are difficult to see. It has also been shown that the southern European practice of marinading game meat in vinegar increases the concentration of lead in the edible tissues, when lead pellets are present (Mateo et al. 2006). Adult Inuit people in arctic Canada showed elevation of blood lead levels, the degree of which was positively correlated with the quantity of hunted waterfowl in the diet (Dewailly et al. 2001). Analysis of stable isotope ratios of lead in blood samples indicates that exposure to ammunition is the main cause of elevated blood lead in indigenous people in Canada (Tsuji et al. 2008).

Waterfowl are not the only game species for which contamination from lead shot is a potential hazard. Liver lead levels in 40% of a sample of Canadian birds of several widely hunted non-waterfowl species (Wild Turkey (*Meleagris gallopavo*), Common Pheasant (*Phasianus colchicus*) and partridge) exceeded health guidelines (Kreager et al. 2008). The main source of lead in these species is thought to be spent lead shot ingested by the birds, having been mistakenly identified as grit. A UK study (Pain et al. 2010) looking at meat from six gamebird species (Red grouse (*Lagopus lagopus*), Mallard, partridge, Common Pheasant, Woodpigeon (*Columba palumbus*) and Woodcock(*Scolopax rusticola*)) found even after visible lead shot had been removed, a high proportion of samples had lead concentrations exceeding the European Union Maximum Level of 100 ppb wet weight. (0.1 mg kg^2 wet weight.) for meat from bovine animals, sheep, pigs and poultry (no level is set for game meat), some by several orders of magnitude. These studies looked primarily at lead shot, which may have a greater risk of secondary poisoning than bullets. However the significant fragmentation observed in lead bullets in shot deer suggests there is likely to be a similar concern for venison (Knott et al. 2010).

ALTERNATIVES TO LEAD BULLETS

The evidence for lead being a problem both for wildlife across wetland and terrestrial ecosystems, as well as humans, is compelling and rapidly developing. However, sports shooting using bullets is a widespread and popular pastime across the globe and lethal control of deer is in some places considered a necessary measure for conservation purposes. Balancing the social, cultural, economic and conservation benefits of these activities against the benefits from removal of lead from systems would be very difficult if not impossible in most situations. It is therefore necessary to carefully assess the efficacy of lead free alternatives to ensure deer shooting using rifles can continue without undue limitations. While steel is the most widely available and used alternative to lead gunshot (as well as, to a lesser degree, tungsten, bismuth and alloys thereof), these substances do not have sufficient density/weight to operate effectively as rifle bullets. The only commercially available lead free bullets consist primarily of copper. In order to be considered an effective alternative to traditional lead bullets, any alternative must address five key areas. The properties of copper bullets in each of these areas will be briefly discussed.

Toxicity

Metallic copper is usually not toxic if ingested and significant deleterious health effects in humans have only been noted following ingestion of massive quantities of the metal (eg. Yelin et al., 1987). Any fragments of copper bullets remaining in, for example, deer carcasses, are therefore highly unlikely to pose any health threat to wildlife or humans. 98.5% of copper ingested is excreted, so minimal build up is possible in living organisms (Kenntner, 2009) limiting the possibility for accumulation of negative effects. In large quantities copper can be toxic to trees, but the quantity contained in a stray bullet would be not be sufficient to cause damage. Copper bullets therefore pose minimal secondary or primary poisoning risk to wildlife, humans or the wider environment.

Safety

Some practitioners have expressed concerns that, because copper bullets do not fragment as much as lead bullets on impact, the risk of ricochet will be greater, particularly if the target is missed. These concerns have led to a moratorium on the use of copper bullets in Germany. While this concern is valid, it has been exaggerated in some quarters, from a single incident in Germany, publicised for political reasons, and was likely due to a lack of health and safety considerations at the site. In practice, any bullet can ricochet under certain conditions and, while impossible to quantify, the increased risk of ricochet from copper bullets appears small and managing that risk should not significantly affect operations. The available evidence shows no increase in reports of accidents caused by ricochet in areas where copper bullets have been widely distributed and used. In a study carried out by the FFU (a German hunting organisation) into hunter attitudes to copper bullets, where 89% of hunters were aware of the relative merits of copper and lead bullets, only 1.3% of respondents mentioned safety concerns (Schuck-Wersig, 2009). There are no published studies showing an increase ricochet risk from use of copper bullets and given their widespread use in some areas (for example by the American army) this suggests use of copper bullets carries no significant additional risk beyond that normally associated with use of any form of ammunition. While some of the concerns expressed are no doubt genuine and it is dangerous to insinuate motivations, these concerns around an increase risk of ricochet compared to lead bullets, are not supported by the evidence.

Lethality

While there are relatively few quantified field trials explicitly designed to compare the lethality of copper and lead bullets, the ballistic properties of the lead and copper bullets of similar design show significant overlap (Hollerman et al. 1990) and do not suggest any significant difference in ballistic performance and hence killing power in field use. The widespread use for deer (and other vertebrate) control work in some areas, notably Arizona and California in America, Hokkaido in Japan and much of Scandinavia, further suggests sufficient lethality in a wide range of situations and for the full range of target species. Indeed, in some situations it has been suggested copper bullets may actually be ballistically superior to lead (Cade, 2007).

A field trial in the UK (Knott et al., 2009), while unscientific and simplistic in its design, does provide a useful insight into the field performance of lead and copper bullets, as its sole function was to compare their efficacy. This field comparison found little difference in killing power for several deer species, although mainly Red Deer (*Cervus elaphus*). Some concern was expressed with use for culling of a smaller species, the Sika (*Cervus nippon*), but these concerns appeared to be corrected for when normal practices at the site were employed.

A similar study, using various calibres and bullet types, carried out in Germany (Spicher, 2009) found 95% of animals (n=260) were killed with a single shot. 48% of beasts did not run and the average distance managed by those that did was just 14 metres. Of the 12 hunters used in the survey, eight (66%) were satisfied copper bullets were as suitable as traditional lead bullets, and four (33%) considered the copper bullets performed better. Only one individual was not satisfied that copper bullets performed as well as lead alternatives.

Overall, there appears to be little or no difference in lethality between copper and lead bullets when used by experienced stalkers.

Availability

Although availability is often a function of demand, it is vital if hunters can reasonably be expected to use any alternative. Availability of copper bullets in different calibres varies greatly between countries, but importantly, the range available from manufacturer's websites and most ammunition dealers, is constantly increasing. Some manufacturers such as Barne's bullets (www.barnesbullets.com) specialise in manufacturing only lead free, copper bullets, and now provide bullets for the full-range of calibres likely to be used for shooting of deer. While local availability may vary, the production of such bullets available for purchase globally, means availability should not be considered a limiting factor for copper bullet use.

Cost

Cost can also be partially a function of demand, with both ammunition producers and dealers likely to charge higher prices for products produced and marketed in smaller quantities, due to economies of scale. However, prices from manufacturers do not appear to differ significantly between equivalent lead and copper bullets. A box of 50 copper bullets for a .270 calibre rifle retail online for $34.64, with an equivalent box of lead bullets costing $36.95 (prices correct as of June 2011). Although there may be significant local variation in prices, these are likely to be influenced by local economies of scale and purchasing patterns. There is no material reason why copper and lead bullet should differ significantly in price and so cost is unlikely to be a long-term limiting factor in copper bullet use.

To Regulate, or Not to Regulate

If it is accepted that use of lead bullets for deer control poses an appreciable health concern, both for humans and wildlife, and that suitable alternatives are available, it is pertinent to consider how best to affect the desired behavioural change, away from continued use of lead bullets. Broadly speaking, there are two possible approaches – voluntary initiatives, based around encouraging and fostering a switch away from lead bullets – and regulation, to force through a switch through legal mechanisms.

Regulations amounting to a partial ban on the use of lead ammunition exist in many countries. However, due to the more established nature of the concern, many of these are focussed towards wetlands. This is reflected in the recognition that steps should be taken to phase out lead ammunition use over wetlands in international agreements, such as the African-Eurasian Waterbird Agreement (AEWA, 1999, 2002, 2008). Regulation affecting lead bullets is much more limited, with only California, Sweden, Denmark and parts of Germany. For a fuller review of international regulation of lead ammunition use see Mateo (2009).

Nor does lead ammunition use necessarily cease, even in areas where it is legislated against. For example, despite lead ammunition use over wetland areas and for shooting certain wetland species being banned in England in 1999 (Environmental Protection (Restriction on Use of Lead Shot (England) Regulations (amended 2002, 2003), a Government commissioned report found 73% of mallard available from game dealers were still being shot with lead (Cromie et al. 2010). The same study also noted the most commonly cited reasons for non-compliance were a belief that the Regulations were unjustified and that they were considered too complex and difficult to comply with.

In practice, any regulation will only be effective if it is effectively enforced and an effective deterrent is applied. The partial bans on lead ammunition use, regulating its use only over certain habitats, or for shooting certain species, most commonly in place across the globe make enforceability very difficult, reflected in the tiny number of successful prosecutions. In the UK, there has only ever been one prosecution for use of lead ammunition (*Quince vs Crown*) and even in this one case, it is likely the offence only came to light because two Mute Swans (*Cygnus olor*), a fully protected species, were shot and the lead ammunition was subsequently discovered. The defendant was fined £100 for use of lead ammunition (plus £500 for illegally killing the two mute swans). Whether or not this would be an effective deterrent is open to debate.

In areas and for ammunition types where legislation has not been imposed, voluntary approaches have been used as an alternative technique to shift behaviours. These can take many forms, but the most common approaches are based around education and economic imperative.

People actively engaged in hunting practices, including deer hunting, consume on average significantly more game meats than the general population (Mintel, 2007). This is perhaps unsurprising given their ready access to the by-products of their hunting activity, even where for example, sporting interest or conservation is the primary motivation for and objective of activities, rather than food production. However, one side-effect is that those actively engaged in use of lead bullets and other ammunitions in game shooting, are one of the groups most likely to suffer form the deleterious health consequences. Education is therefore a potentially powerful tool in shifting behaviours. Provision of information regarding the potential risks and health concerns associated with lead poisoning from ammunition sources, can provide an educational "push" away from use of lead ammunition.

If education can provide a "push" away from use of lead ammunition, economic incentification can provide a "pull" towards alternatives. As already discussed, there is little cost difference between lead bullets and copper alternatives and any differences are likely to disappear as copper bullets become more widely used in any given area, due to economies of scale. While this is helpful in that it does not provide a significant block to switching form lead to copper bullets, it also does not actively encourage such a shift. Mechanisms to make copper bullets more financially attractive to purchasers than their lead equivalents can be applied, either by artificially raising the price of lead bullets, or by depressing the price of alternatives.

These "push" (education) and "pull" (economic incentification) techniques are likely to be most effective when combined in a given area, in order to shift the balance of costs and benefits (perceived and real) towards favouring a behavioural shift away from lead ammunition. These approaches have been used in areas of Arizona, where information has been provided to hunters around fragmentation of lead bullets and the associated health risks, including use of a portable x-ray machine to graphically illustrate fragmentation in hunters' own deer. At the same time, an economic incentive for switching has been offered in the form of a free box of copper bullets, provided by the Arizona Game and Fish Department for hunters willing to make the change. This combination of techniques (along with others) has shown significant success, with an estimated 80% of hunters using copper bullets within the target area (Sieg et al. 2009).

However, there are limits to the potential success of such an approach. Many hunters and groups are deeply traditional and behavioural change can therefore be slow, uneven, or otherwise unsusceptible to voluntary approaches. A study into the attitudes of German hunters to copper bullets found 27% had no reservations about switching to copper bullets and 67% were prepared to switch, after taking into account all available evidence and experience (Schuck-Wersig, 2009). This is interesting as it suggests that while a majority of hunters would be susceptible to education based voluntary approaches, a significant minority (in this case 33%) would be unwilling or unlikely to change their practices, even once all available evidence has been presented. It is unknown what effect economic incentives would have on this figure. Even in Arizona, while 80% behavioural change is an impressive achievement, this still means a significant minority (20%) continue to use lead ammunition. Even this reduced level of lead exposure in the environment is a conservation concern, as illustrated by the continued of mortality of California Condors due to lead poisoning from ammunition sources (Peregrine Fund, press release May 2011).

Voluntary approaches are generally considered preferable if they can deliver the desired outcome, as they reduce the regulatory burden, both for hunters and regulatory bodies. Their ability to do so in the case of encouraging a switch away from lead bullet use will depend on several factors, such as the required reduction in lead use in the target area, who delivers the justification for change, and how the message is delivered. Success is most likely where hunting organisations themselves deliver the messaging, as this is likely to minimise the risk of alienating the sector whose behaviour is desired to change. In deciding whether voluntary or regulatory approaches are most appropriate in a given situation, a series of variables must be considered. For example, voluntary approaches may be most suitable where the hunting community is considered open to behavioural change, hunting organisations support the shift, sufficient resource is available to provide effective education and economic incentives (although note the cost of this should be over the short-term only, in contrast to the smaller but prolonged cost of enforcing regulation), a total elimination of lead use is not required and there is no requirement for an immediate change. If these conditions are not met, voluntary approaches are likely to be ineffective and regulation will be required to achieve the desired reduction in lead exposure from ammunition sources.

CONCLUSION

There is a significant and expanding evidence based showing that lead from ammunition sources, including bullets used for deer culling, poses a significant risk to wildlife through secondary poisoning. Scavengers are most vulnerable and a severe conservation problem can be caused in small and vulnerable populations. There is also growing evidence of deleterious impacts on human health through ingestion of ammunition fragments not removed by the standard butchering process. This is a particular concern for high risk groups such as young children, pregnant women and high consumers of game meat. While responses to lead poisoning of waterfowl from ammunition of use over wetlands have been enacted or at least recommended in many areas (although enforcement remains a significant concern), responses to terrestrial and human health risks have lagged in most countries. Given the risks and the availability of apparently suitable alternatives, a general switch away from use of lead bullets for deer culling would be beneficial. This could either be achieved through voluntary measures of regulation. While the former is less likely to alienate sections of the hunting community and may be preferable in some circumstances, it is unlikely to deliver total cessation of lead ammunition use over the short to medium term and may never do so if sufficient resource is not available to support behavioural change, or if the hunting community in the area is resistant to change. Regulation has the ability to achieve a rapid and more complete shift in away from lead ammunition, but can lead to problems of its own, including the difficulty and ongoing cost of effective enforcement and importantly, the risk of alienating key stakeholders.

The details of the most appropriate response will vary between areas and groups and should be discussed and decided upon by all relevant stakeholders, including hunters, conservationists, consumers, food providers and suppliers and Governments. However, the issue of lead poisoning from ammunition sources is significant and further efforts should be invested by all interest groups in order to foster a permanent behavioural shift away from use of lead ammunition.

REFERENCES

African-Eurasian Migratory Waterbirds (AEWA) (1999) Resolution 1.14 Phasing out of lead shot in wetlands, *First Meeting of the Parties to the Agreement on the Conservation of African-Eurasian Migratory Waterbirds (AEWA)*, 6-9 November 1999, Cape Town, South Africa.

African-Eurasian Migratory Waterbirds (AEWA) (2002) Resolution 2.2 Phasing out of lead shot for hunting in wetlands, *Second Meeting of the Parties to the Agreement on the Conservation of African-Eurasian Migratory Waterbirds (AEWA)*, 25-27 September 2002, Bonn, Germany.

African-Eurasian Migratory Waterbirds (AEWA) (2008) Resolution 4.1 Phasing out of lead shot for hunting in wetlands, *Fourth Meeting of the Parties to the Agreement on the Conservation of African-Eurasian Migratory Waterbirds (AEWA)*, 15-19 September 2008, Antananarivo, Madagascar.

Bonet, A, Olivares, C, Pico, ML and Sales, S (1995) L'acumulació de perdigons de plom al Parc Natural del Fondó d'Elx (Alacant): distribució espacial i propostes d'actuació. Butlletí de la Instució Catalana d'Història, *Natural* 63; 149-166.

Cade, TJ (2007) Exposure of California condors to lead from spent ammunition, *Journal of Wildlife Management* 71 (7); 2125-2133.

Canfield, RL, Henderson, CR, Cory-Slechta, DA, Cox, C, Jusko, TA and Lanphear, BP (2003) Intellectual impairment in children with blood lead concentrations below 10 mu g per deciliter. *New England Journal of Medicine* 348; 1517-1526.

Cerradelo , S, Munoz, E, To-Figueras, J, Mateo, R and Guitart, R (1992) Intoxicacion por ingestion de perdigones se plomo en dos aguilas reales, *Donana Acta Vertebrate* 19; 122-127.

Chandramouli, L, Steer, CD, Ellis, M and Emond, AM (2009) Effects of early childhood lead exposure on academic performance and behaviour of school age children, *Archives of Disease in Childhood* 94; 844-848.

Chesley J, Reinthal, P, Parish, C, Sullivan, K and Sieg R (2009) Evidence for the source of lead contamination within the California Condor. In: Watson, RTM, Fuller, M, Pokras, M and Hunt, WG (Eds.) *Ingestion of spent lead ammunition: implications for wildlife and humans*, The Peregrine Fund, Boise, Idaho, USA.

Cromie, RL, Loram, A, Hurst, L, O'Brien, M, Newth, J, Brown, MJ and Harradine, JP (2010) Compliance with the Environmental Protection (Restrictions on Use of Lead Shot)(England) Regulations 1999, Report to Defra, Bristol, UK.

Dewailly, E, Ayotte, P, Bruneau, S, Lebel, G, Levallois, P and Weber, JP (2001) Exposure of the Inuit population of Nunavik (Arctic Quebec) to lead and mercury. *Archives of Environmental Health* 56; 350-357.

Dobrowolska A and Melosik M (2008) Bullet-derived lead in tissues of the wild boar (*Sus scrofa*) and red deer (*Cervus elaphus*), *European Journal of Wildlife Research* 54; 231-235.

EFSA (2010) EFSA Panel on Contaminants in the Food Chain (CONTAM): Scientific opinion on lead in food, *EFSA Journal* 8 (4); 1570-1717.

Green, RE, Hunt, WG, Parish, CN and Newton, I (2009) Modelling blood lead concentration and exposure in free-ranging California condors in Arizona and Utah, *In* Watson, RTM, Fuller, M, Pokras, M and Hunt WG (Eds.) *Ingestion of Spent Lead Ammunition: Implications for Wildlife and Humans*, The Peregrine Fund, Boise, Idaho, USA.

Hollerman, JJ, Fackler, ML, Coldwell, DM and Ben-Menachem, Y (1990) Gunshot wounds: Bullets, ballistics and mechanisms of injury, *American Journal of Roentgenology* 155; 685-690.

Hunt, WG, Burnham, W, Parish, CN, Burnham, KK, Mutch, B and Oaks, JL (2006) Bullet fragments in deer remains: Implications for lead exposure in avian scavengers, *Wildlife Society Bulletin* 34; 167-170.

Hunt, WG, Watson, RT, Oaks, JL, Parish, CN, Burnham, KK and Tucker, RL (2009) Lead bullet fragments in packaged venison: potential for human dietary exposure, *In* Watson, RTM, Fuller, M, Pokras, M and Hunt WG (Eds.) *Ingestion of Spent Lead Ammunition: Implications for Wildlife and Humans*, The Peregrine Fund, Boise, Idaho, USA.

Iqbal, S, Blumenthal, W, Kennedy, C, Yip, FY, Pickard, S, Flanders, WD, Loringer, K, Kruger, K, Caldwell, KL and Brown, MJ (2009) Hunting with lead: Association between blood lead levels and wild game consumption, *Environmental Research* 109; 952-959.

Johansen, P, Asmund, G and Riget, F (2004) High human exposure to lead through consumption of birds hunted with lead shot. *Environmental Pollution* 127; 125-129.

Kenntner, N (2009) *Toxicokinetic and toxicity of lead and alternative hunting ammunition.* Presentation to '*Lead Intoxication of white-tailed sea eagles: causes and potential solutions*' conference, Berlin, Germany.

Kenntner, N, Crettenand, Y, Funfstuck, HJ, Janovsky, M and Tataruch, F (2007) Lead poisoning and heavy metal exposure of golden eagles (Aquila chrysaetos) from the European Alps, *Journal of Ornithology* 148; 173-177.

Knott, J, Gilbert, J, Green, RE and Hoccom, DG (2009) Comparison of the lethality of lead and copper bullets in deer control operations to reduce incidental lead poisoning; field trials in England and Scotland, *Conservation Evidence* 6; 71-78.

Knott, J, Gilbert, J, Hoccom, DG and Green, RE (2010) Implications for wildlife and humans of dietary exposure to lead from fragments of lead rifle bullets in deer shot in the UK, *Science of the Total Environment* 409; 95-99.

Kreager, N, Wainman, BC, Jayasinghe, RK and Tsuji, LJS (2008) Lead pellet ingestion and liver-lead concentrations in upland game birds from southern Ontario, Canada. *Archives Of Environmental Contamination And Toxicology* 54; 331-336.

Krone O (2009) Causes of death in white-tailed sea eagles and sources of lead poisoning. Presentation to '*Lead Intoxication of white-tailed sea eagles: causes and potential solutions*' conference, Berlin, Germany.

Krone, O, Langgemach, T, Sommer, P and Kenntner, N (2003) Causes of mortality in white-tailed sea eagles from Germany. In: Helander, B, Marquiss, M and Bowerman, W (Eds.) Sea Eagle 2000, Proceedings from an international conference at Bjorko, Sweden, 13-17 September 2000, *Swedish Society for Nature Conservation/SNF and Atta. 45 Tryckeri AB*, Stockholm.

Lanphear BP, Hornung R, Khoury J, Yolton K, Baghurst P, Bellinger D, Canfield RL, Dietrich KN, Bornschein R, Greene T, Rothenberg SJ, Needleman HL, Schnaas L,Wasserman G, Graziano J, Roberts R (2005) Low-level lead exposure and children's intellectual function: an international pooled analysis, *Environmental Health Perspectives*, 113; 894-899.

Martin PA, Campbell D and Scheuhammer A (2003) Lead exposure in terrestrial foraging raptors in southern Ontario, 1999-2001, *International Association for Great Lakes Research Conference Program and Abstracts* 44, IAGLR, Ann Arbor; 269.

Mateo, R (2009) Lead poisoning in wild birds in Europe and the rregulations adopted by different countries, *In* Watson, RTM, Fuller, M, Pokras, M and Hunt WG (Eds.) *Ingestion of Spent Lead Ammunition: Implications for Wildlife and Humans*, The Peregrine Fund, Boise, Idaho.

Mateo, R, Belliure, JC, Aguilar-Serrano, JM and Guitart, R (1998) High prevalences of lead poisoning in wintering waterfowl in Spain, *Archives of Environmental Contamination and Toxicology* 35:342-347.

Mateo, R, Green, AJ, Lefranc, H, Baos, R and Figuerola, J (2007) Lead poisoning in wild birds from southern Spain: A comparative study of wetland areas and species affected, and trends over time, *Ecotoxicology and Environmental Safety* 66; 119-126.

Mateo, R, Rodriguez-de la Cruz, M, Vidal, D, Reglero, M, and Camero, P, (2006) Transfer of lead from shot pellets to game meat during cooking. *Science of The Total Environment* 372; 480-485.

Mateo, R., Taggart, M., Meharg, A.A., 2003. Lead and arsenic in bones of birds of prey from Spain. *Environmental Pollution*, 126; 107–114.

Miller, MJR, Wayland, ME and Bortolotti, GR (2002) Lead exposure and poisoning in diurnal raptors: a global perspective. In: Yosef, RM, Miller, ML and Pepler, D (Eds.) Raptors in the new millennium, *Proceedings of the Joint Meeting of the Raptor Research Foundation and the World Working Group on Birds of Prey and Owls*, Eilat, Israel, 2-8 April 2000; International Birding and Research Centre, Eilat, 224-245.

Mintel (2007) *Game and exotic meat in the UK*, Mintel Commissioned Report.

Mudge, GP (1983) The incidence and significance of ingested lead pellet poisoning in British wildfowl, *Biological Conservation* 27; 333-372.

Pain, DJ (1991) Lead shot densities and settlement rates in Camargue marshes, France, *Biological Conservation* 57; 273-286.

Pain, DJ (1992) Lead poisoning in waterfowl: a review. In: Pain, DJ (Ed.), Lead Poisoning in Waterfowl, *IWRB Spec. Pub. 16*, Slimbridge; 7-13.

Pain, DJ, Carter, I, Sainsbury, AW, Shore, RF, Eden, P, Taggart, MA, Konstantinos, S, Walker, LA, Meharg, AA and Raab, A (2007) Lead contamination and associated disease in captive and reintroduced red kites *Milvus milvus* in England. *Science of the Total Environment* 376; 116-127.

Pain, DJ, Cromie, RL, Newth, J, Brown, MJ, Crutcher, E, Hardman, P, Hurst, L, Mateo, R, Meharg, AA, Moran, AC,, Raab, A, Taggart, MA and Green, RE (2010) Potential Hazard to Human Health from Exposure to Fragments of Lead Bullets and Shot in the Tissues of Game Animals, PLoS ONE 5 (4) 1-17.

Pain, DJ, Fisher, IJ and Thomas, VG (2009) A global update on lead poisoning in terrestrial birds from ammunition sources, *In* Watson, RTM, Fuller, M, Pokras, M and Hunt WG (Eds.) *Ingestion of Spent Lead Ammunition: Implications for Wildlife and Humans*, The Peregrine Fund, Boise, Idaho.

Parish, CN, Hunt, WG, Feltes, E, Seig, R and Orr, K (2009) Lead exposure among a reintroduced population of California condors in northern Arizona and southern Utah, *In* Watson, RTM, Fuller, M, Pokras, M and Hunt WG (Eds.) *Ingestion of Spent Lead Ammunition: Implications for Wildlife and Humans*, The Peregrine Fund, Boise, Idaho.

Peterson, BD and Meltofte, H (1979) Forekomst af blyhagl i vestjyske vådområder samt i kråsen hos danske ænder, *Dansk Ornitologisk Forenings Tidsskrift* 73; 257-264.

Platt, JB (1976) Bald eagles wintering in the Utah desert, *American Birds* 30; 783-788.

Potts, GR (2005) Incidence of ingested lead gunshot in wild grey partridges (*Perdix perdix*) from the UK, *European Journal of Wildlife Research*; 51, 31-34.

Schuck-Wersig, P (2009) Awareness of lead intoxication in birds of prey: results of a German hunter survey, *Presentation to 'Lead intoxication of white-tailed sea eagles: causes and potential solutions' conference,* Berlin, Germany.

Sieg, R, Sullivan, KA and Parish, CN (2009) Voluntary lead reduction efforts within the northern Arizona range of the California condor, *In* Watson, RTM, Fuller, M, Pokras, M and Hunt WG (Eds.) *Ingestion of Spent Lead Ammunition: Implications for Wildlife and Humans*, The Peregrine Fund, Boise, Idaho.

Spicher, V (2009) Experience with lead-free bullets – a field report, Presentation to *'Lead intoxication of white-tailed sea eagles: causes and potential solutions'* conference, Berlin, Germany.

Street, M (1983) The assessment of mortality resulting from the ingestion of spent lead shot by Mallard wintering in south-east England, *In* Proceedings of the XI Congress of the International Union of Game Biologists, Trujillo, Spain.

Thomas, GJ (1982) Lead poisoning in waterfowl, *In* Managing Wetlands and Their Birds: A Manual of Wetland and Waterfowl Management, *International Waterfowl and Wetlands Research Bureau*, Slimbridge, UK, 260-268.

Tsuji, LJS, Wainman, BC, Martin, ID, Sutherland, C, Weber, JP, Dumas, P and Nieboer, E (2008) The identification of lead ammunition as a source of lead exposure in First Nations: The use of lead isotope ratios. *Science of The Total Environment* 393, 291-298.

Watson R (2009) Lead poisoning in the Californian condor in the USA, Presentation to *'Lead intoxication of white-tailed sea eagles: causes and potential solutions'* conference, Berlin, Germany.

Yelin, G, Taff, ML and Sadowski, GE (1987) Copper toxicity following massive ingestion of coins, *American Journal of Forensic Medical Pathology* 8(1); 78-85.

In: Deer: Habitat, Behavior and Conservation ISBN 978-1-62100-676-3
Editors: A. A. Cahler and J. P. Marsten

Chapter 5

Two Genetically Distinct, Endemic Subspecies of Deer from Korea (Chinese Water Deer *Hydropotes inermis argyropus* and Siberian Roe Deer *Capreolus pygargus ochracea*), Examined with Nuclear DNA *IRBP* and Mitochondrial DNA Cytochrome *b* and Control Region Sequences, and Their Conservation in Korea

Hung Sun Koh,[1*] Damdingiin Bayarlkhagva,[2] Kyung Hee Jang,[1] Byong Guk Yang,[3] Bae Keun Lee,[1] and Jang Geun Oh[4]

[1]Department of Biology, Chungbuk University, Cheongju, Korea
[2]Department of Molecular Biology, National University of Mongolia, Ulaanbaatar, Mongolia
[3]National Institute of Environmental Research, Incheon, Korea
[4]Research Center of Environment, Jeju, Korea

* Corresponding author. Tel; +82 43 2612292: Fax; +82 43 2532291. E-mail: syskoss@chungbuk.ac.kr.

ABSTRACT

We examined genetic distinctness of two endemic subspecies of deer in Korea [Chinese water deer (*Hydropotes inermis argyropus*) in Korean peninsula and Siberian roe deer (*Capreolus pygargus ochracea*) in Korean peninsula and Korean Jeju Island] by nuclear *IRBP* and mitochondrial cytochrome *b* and control region sequence analyses of two species in Asia. Two sympatric phylogroups of Chinese water deer in Korea, previously recognized from our cytochrome *b* and control region sequence analysis, were indistinguishable in this *IRBP* sequence analysis. On the other hand, genetic distinctness of the Siberian roe deer in Jeju Island, previously revealed from our control region sequence analysis, was confirmed from present *IRBP* analysis. We concluded that Jeju Island roe deer is an allopatric phylogroup, which differs from other populations in Asia, including Korean peninsula, northeastern China, far-eastern Russia, Mongolia, and western Siberian Kurgan. Each endemic subspecies of deer in Korea is genetically distinct to be considered as evolutionary significant unit (ESU), and each subspecies is unique in their distributional aspect as well. However, conservation status of two deer in Korea is in striking contrast. The former is abundant, and is regarded as a pest, game mammal, whereas the latter was rare and is treated as a protected mammal. In order to determine protection tactics for two subspecies of deer in Korea and other deer from other regions as well, we recommend adoption of a criterion, i.e., evaluation of distributional and evolutionary uniqueness in global scale, to upgrade into threatened category, when population size in a certain local area is large enough to fit the IUCN guideline for the least concern category.

Keywords: *Hydropotes inermis*, *Capreolus pygargus*, *IRBP*, Korea, conservation genetics

INTRODUCTION

The Korean peninsula (220,847 km^2) in East Asia extends southwards from northeastern China and far-eastern Russia: northern boundaries are formed naturally by Amnok/Yalu river, Baegdu/Baitou Mountain (the main peak of Changbai Mountains in northeastern China), and Duman/Tumen river.

Jeju Island (1,846 km^2) in Korea is one of major islands in the Yellow sea, and its location is about 100 km off the southern coasts of the Korean peninsula.

The mammals of the Korean peninsula are not diverse compared with other parts of East Asia because of the small area and limitation of free dispersal by the natural barriers (Changbai Mountains, including Baegdu Mountain, and the rivers of Amnok and Duman): 185 species of 11 orders in East Asia, 85 species of seven orders in northern part of Korea, and 44 species of seven orders in southern part of Korea (Corbet, 1978): three species of Cervidae (*Cervus nippon*, *Hydropotes inermis*, and *Capreolus pygargus*) distribute in Korea, and one species of Cervidae (*C. pygargus*) in Jeju Island.

The conservation of biodiversity has three levels, i.e., genetic, species, and community/ecosystem, and the genetic diversity means the genetic variation found within each species (Primack, 1995). Genetic diversity can be measured at a number of different levels, i.e., measurable characters, visible direct effects of deleterious alleles, protein variation, and direct measurement of DNA sequence variation (Frankham et al., 2004), and genetic diversity estimates are usually higher when based on microsatellite data compared with dominant data (Freeland, 2005).

The importance of examining both nuclear and mitochondrial loci when attempting to elucidate patterns of genetic structure was noted (Chappell et al., 2004). However, the evolutionary rates of nuclear genes varies slower than those of mitochondrial DNA (mtDNA) sequences (Steppan et al., 2005), and among mtDNA sites cytochrome *b* genes are highly conserved, whereas control region is the most divergent regions (Lopez et al., 1997).

The sika deer (*C. nippon*) distributes in east Siberia, northern and southern China, Taiwan, Korea, and Japan, and the sika deer from southeastern Siberia, Korea, and China is named as *C. n. hortulorum* (Woon, 1967). The sika deer (*C. n. hortulorum*) from Korea is listed as class I, critically threatened mammal (Korean Ministry of Environment, 2005), although it is extinct in the wild. From the comparison with cytochrome *b* sequence of the sika deer from deer farms in northeastern China and Korea, Koh et al. (2010) concluded that there is no sika deer of subspecies *hortulorum* in deer farms of Korea.

The Chinese water deer (*Hydropotes inermis*) is a single species of genus *Hydropotes*, and it was formerly widespread in China (Guangxi to Liaoning) and Korea, with two parapatric subspecies, i.e., *H. i. inermis* from China and *H. i. argyropus* from Korea (Grubb, 2005). At present, the Chinese water deer in China (*H. i. inermis*) survives in the lower Yangtze Basin, west to Hupeh (Corbet, 1978). The Chinese water deer in Korea (*H. i. argyropus*) is abundant and is a pest, game mammal, although Koh et al. (2009) recognized two, sympatric phylogroups within the Chinese water deer from Korea by cytochrome *b* and control region sequence analyses.

The Siberian roe deer (*C. pygargus*) distributes in east European Russia, Siberia, and China, and it is classified into two subspecies: *C. p. pygargus* from the Volga area, the Urals, western and part of eastern Siberia (not farther than Baikal), the Altai, and Yakutia and *C. p. tianschanicus* from the Tien Shan, Mongolia, northern and central China (Xinjiang, eastern Qinghai, western Sichuan, and northeastern China), Korea, and far-eastern Russia (Danilkin, 1996), although eight subspecies, including *C. p. ochracea* from Korea and *mantschuricus* (= *bedfordi*) from northeastern China, were recognized previously (Grubb, 2005). The Siberian roe deer in Korean Jeju Island is also *C. p. ochracea* (Woon, 1967) and is protected by the Jeju provincial government. The Jeju Island roe deer was distinct from the Siberian roe deer of western Siberian Kurgan and far-eastern Russian Amur in their mtDNA control region sequences (Koh and Randi, 2001), and they confirmed the subspecies name of *C. p. ochracea.*

In this paper, we examined genetic distinctness of two endemic subspecies of deer in Korea [the Chinese water deer (*Hydropotes inermis argyropus*) from the Korean peninsula and the Siberian roe deer (*Capreolus pygargus ochracea*) from the Korean peninsula and Korean Jeju island] by nuclear *IRBP* and mitochondrial cytochrome *b* and control region sequence analyses of two species in Asia, and we discussed their proper conservation tactics.

MATERIALS AND METHODS

All 30 specimens of the Chinese water deer (*H. inermis*) in China and Korea were used for *IRBP* sequence analysis. In China, seven specimens from two locations (Dafeng, Jiangsu, HCD09-HCD10; and Zhoushan, Zhejiang, HCZ03-HCZ04 and HCZ06-HCZ08) were used, and in Korea, 23 specimens of five locations (Mt. Seolak, HKS03 and HKS05; Mt. Weolak, HKW02-HKW08; Boeun, HKB01-HKB02, HKB04, HKB07-HKB08; Cheongju, KC02-HKC09; and Mt. Jiri, HKJ08) were used. Geographic locations of these seven sites are shown in Figure 1 (6, Dafeng; 7, Zhoushan, 8, Mt. Seolak; 9, Mt. Weolak; 10, Boeun; 12, Cheongju; and 13, Mt. Jiri).

All 21 specimens of the Siberian roe deer (*C. p. tianschanicus*) in five regions of East Asia were used for *IRBP* sequence analysis [Mongolia (Ulaanbaatar, CMU01-CMU04), far-eastern Russia (Vladivostok, CRV01-CRV02), northeastern China (Yichun, CCY01-CCY04; Harbin, CCH01; and

Shenyang, CCS01), Korea (Mt. Seolak, CKS01; Mt. Weolak, CKW01-CKW02; and Mt. Songri, CKs01-CKs03), and Korean Jeju Island (CKJ01-CKJ03)]. Collection sites are shown in Figure 1 (1, Ulaanbaatar; 2. Vladivostok; 3, Yichun; 4, Harbin; 5, Shenyang; 8, Mt. Seolak; 9, Mt. Weolak; 11, Mt. Songri; and 14, the Jeju Island).

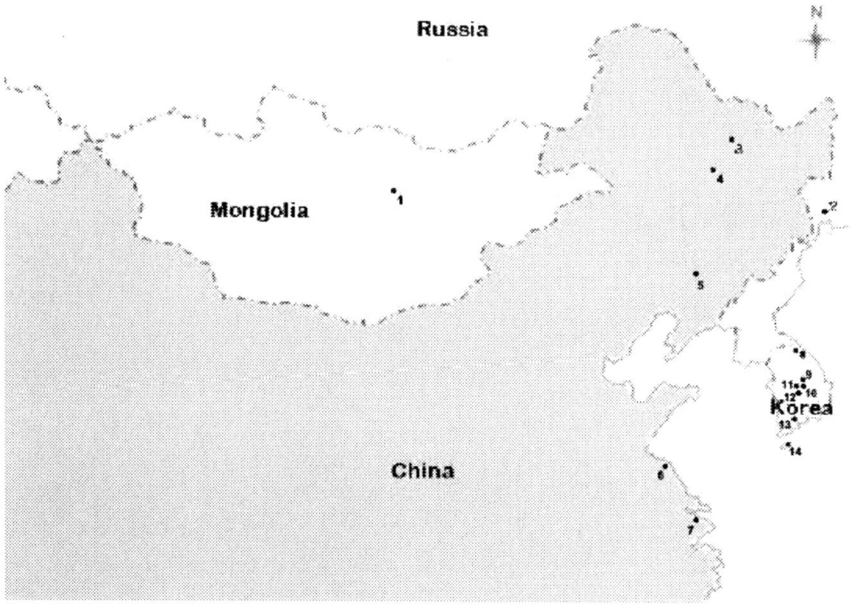

Figure 1. Collection sites of 51 specimens of two species of deer from Korea (*Hydropotes inermis* and *Capreolus pygargus*). The locations of seven sites in *H. inermis* are (6, Dafeng; 7, Zhoushan, 8, Mt. Seolak; 9, Mt. Weolak; 10, Boeun; 12, Cheongju; and 13, Mt. Jiri), and nine sites for *C. pygargus* are (1, Ulaanbaatar; 2, Vladivostok; 3, Yichun; 4, Harbin; 5, Shenyang; 8, Mt. Seolak; 9, Mt. Weolak; 11, Mt. Songri; and 14, Jeju Island).

Muscle samples were stored in a deep-freezer, and total cellular DNA was extracted from muscle samples with DNA extraction kit from Bioneer Co. (Chungbuk, Korea). The *IRBP* gene was PCR-amplified using primers IRBP217 and IRBP1531, designed by Stanhope et al. (1992). PCR thermal cycle was as follows: 94°C for five min; 94°C for 45sec, 60°C for 45 sec, and 72°C for one min (30 cycles); and 72°C for five min. Excess primers and unincorporated nucleotides were eliminated by purifying the amplified product with DNA PrepMate kit from Bioneer Co. (Chungbuk, Korea). The purified

PCR products were sequenced using an automated DNA Sequencer (model 377, Perkin Elmer) from Macrogen Co. (Seoul, Korea).

Tamura-Nei nucleotide distances (Tamura and Nei, 1993) were calculated, and phylogenetic trees were constructed by neighbor-joining methods with 1,000 bootstrap replications by MEGA (version 4.0). *Sus scrofa* (U48588) was used for outgroup.

In addition, our previous results on two Korean deer species (control region sequence analysis of the Siberian roe deer (Koh et al., 2001) and cytochrome *b* and control region sequence analyses of the Chinese water deer (Koh et al., 2009) were used to discuss with present results of *IRBP* analyses.

RESULTS

Chinese Water Deer

The *IRBP* gene sequences of 1,152 bp were obtained from 30 specimens in China and Korea, and four alleles (HA 1, HA 2, HA 3, and HA 4) were revealed, as shown in Figure 2: HA 1 from nine specimens of four locations in Korea (Mt. Weolak, HKW06-HKW08; Boeun, HKB01-HKB02 and HKB08; Cheongju, HKC02-HKC03, and Mt. Jiri, HKJ08), HA 2 from 14 specimens of four locations in Korea and three specimens of one location in China (Mt. Seolak, HKS03 and HKS05; Mt. Weolak, HKW02-HKW5; Boeun, HKB04 and HKB07; Cheongju, HKC04-HKC09; and Zhoushan, HCZ03, HCZ06, and HCZ08), HA 3 from two specimens of one location from China (Zhoushan, HCZ04 and HCZ07), and HA 4 from two specimens of one location in China (Dafeng, HCD09-HCD10)].

Three alleles (HA 2, HA 3, and HA 4) were found from China, and two alleles (HA 1 and HA 2) were recognized from Korea: HA 2 was common from China and Korea. In addition, three sites (site no. 155, 191, and 601) were variable, and the sequences of the variable sites in the four alleles were GCT, GCG, ACG, and GAG, respectively. In addition, as shown in Figure 2, the specimens of *H. inermis* (four alleles) were not distinguishable into distinct subgroups in their *IRBP* sequences (the maximum Tamura-Nei nucleotide distance among the four alleles was 0.17%).

Siberian Roe Deer

The *IRBP* gene sequences of 1,229 bp were obtained from 21 specimens of five regions in East Asia (Mongolia, far-eastern Russia, northeastern China, the Korean peninsula, and Korean Jeju Island), and two alleles (CA 1 and CA 2) were found, as shown in Figure 3. CA 1 was from 18 specimens of four regions (Mongolia (Ulaanbaatar, CMU01-CMU04), far-eastern Russia (Vladivostok, CRV01-CRV02), northeastern China (Yichun, CCY01-CCY04; Harbin, CCH01; and Shenyang, CCS01), and Korea (Mt. Seolak, CKS01; Mt. Weolak, CKW01-CKW02; and Mt. Songri, CKs01-CKs03), and CA 2, from three specimens of one region (Korean Jeju Island, CKJ01-CKJ03). The Siberian roe deer from East Asia was composed of two subgroups (one subgroup of CA 2 from Korean Jeju Island and the other subgroup of CA 1 from the Korean peninsula, northeastern China, far-eastern Russia, and Mongolia). The four sites (no. 563, 626, 690, and 909) were variable, and the sequences of the four variable sites in the two alleles were CCGG and TTAA, respectively. In addition, as shown in Figure 3, Tamura-Nei nucleotide distance between two subgroups, i.e., two alleles, was 0.33%.

DISCUSSION

Chinese Water Deer

The different levels of polymorphism found in vertebrate mtDNAs vs. nuclear genes are attributed to: (1) higher mutation rates of mtDNA, compared to single-copy nuclear DNA, and (2) short-term mtDNA dynamics, which are more strongly influenced by population processes such as genetic drift due to maternal transmission (Chapman, 1989). The Yellow-Sea separating the Korean peninsula and central China was entirely land during the last glacier, about 20,000-10,000 years ago (Chung, 2007), and the Chinese water deer was formerly widespread from China (Guangxi to Liaoning) to the Korean peninsula; but, at present, the Chinese water deer from China (*H. i. inermis*) survives in the lower Yangtze Basin and is listed as near threatened in the IUCN Red List (Grubb, 2005), although the Chinese water deer from (*H. i. argyropus*) is abundant in Korea, and it is a pest, game mammal.

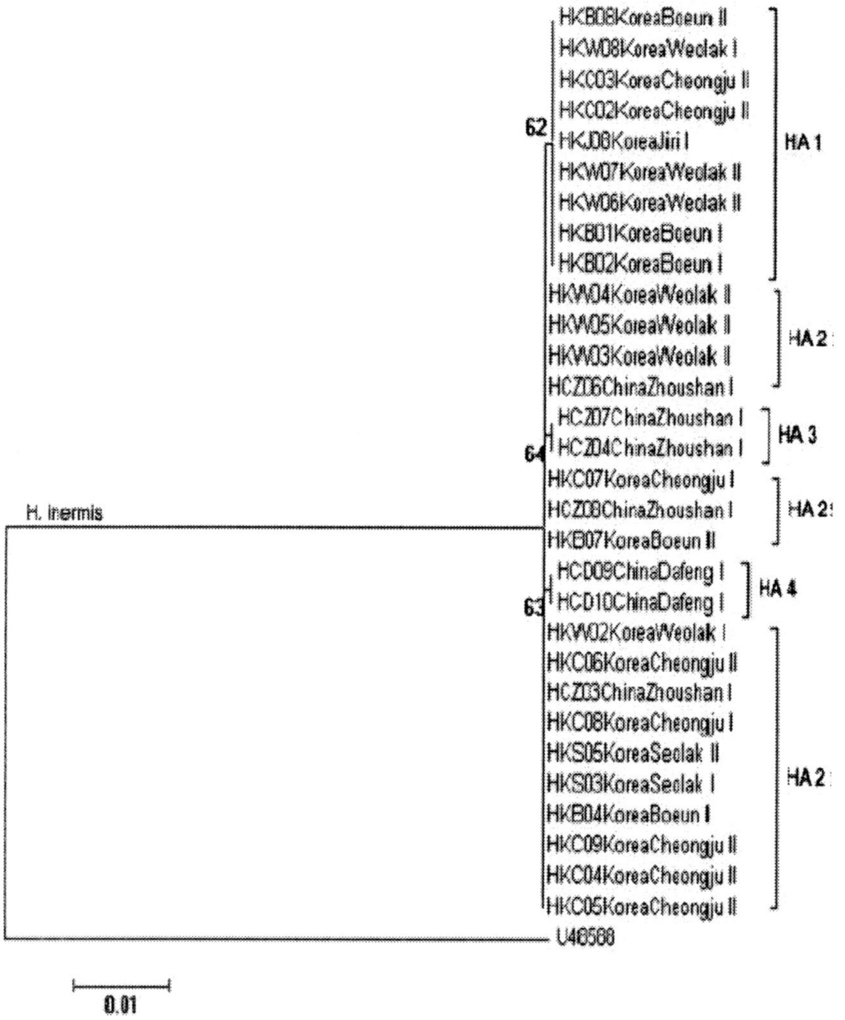

Figure 2. Phylogenetic relationships among 30 specimens, recognized as four *IRBP* alleles (HA 1, HA 2, HA 3, and HA 4), of the Chinese water deer (*Hydropotes inermis*) from China and Korea. The location of each specimen and the specimen of each allele are given in MandM and Results, respectively. Roman numerals I and II, following specimen number and location, indicate two clades based on cytochrome *b* and control region sequences of *H. inermis* by Koh et al. (2009). The neighbor-joining tree was constructed with 1,000 bootstrapped replications, and bootstrap values greater than 50% are reported at the internodes. *Sus scrofa* (U48588) was used for outgroup.

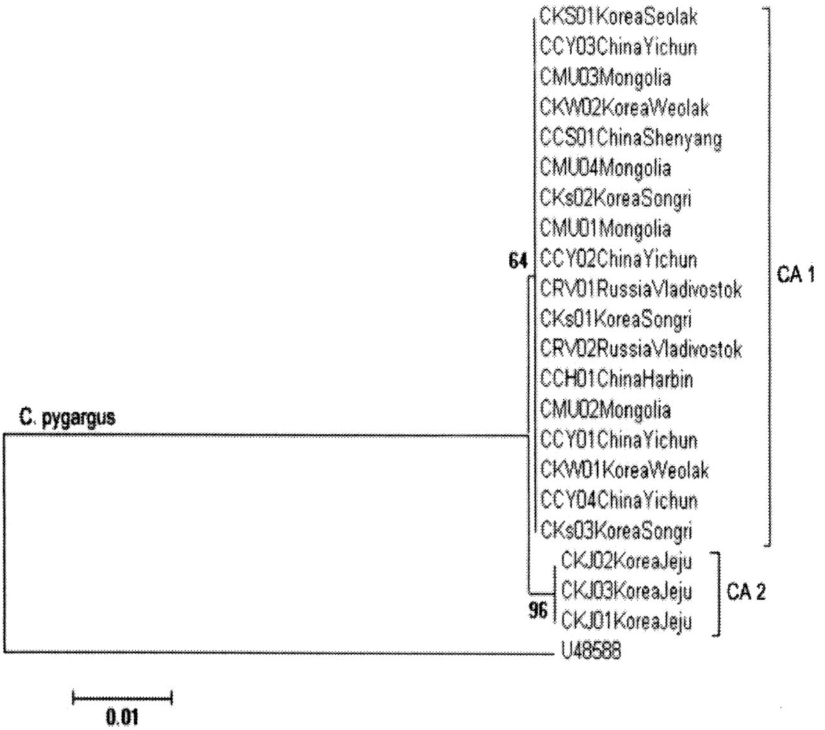

Figure 3. Phylogenetic relationships among 21 specimens, recognized as two subgroups of two *IRBP* alleles (CA1 and CA2), of the Siberian roe deer (*Capreolus pygargus*) from East Asia. The location of each specimen and the specimen of each allele are given in MandM and Results, respectively. The neighbor-joining tree was constructed with 1,000 bootstrapped replications, and bootstrap values greater than 50% are reported at the internodes. *Sus scrofa* (U48588) was used for outgroup.

Koh et al. (2009) found that the Chinese water deer (*H. inermis*) is composed of two sympatric phylogroups (phylogroup I, a major clade from China and Korea; and phylogroup II, a minor clade from Korea) with average Tamura-Nei nucleotide distances of 2.1% in control region and 1.3% in cytochrome *b* gene sequences. In this study with the *IRBP* gene sequences of 1,152 bp, as shown in Figure 2, the four alleles in *H. inermis* [three alleles (HA 2, HA 3, and HA 4) from China and two alleles (HA 1 and HA 2) from Korea] were not differentiated into distinct subgroups (the maximum nucleotide distance among the four alleles was 0.17%).

In addition, as also shown in Figure 2, the two phylogroups (I and II), resulted from cytochrome *b* and control region sequences, were not distinguishable in the *IRBP* sequences: allele HA 1 was composed of nine specimens, which belong to either phylogroup I (4 specimens of KW08, KB01, KB02, and KJ08) or phylogroup II (5 specimens of KW06, HKW07, HKB08, HKC02, and HKC03), and allele HA 2 consisted of 17 specimens of either phylogroup I (8 specimens of HKS03, HKB04, HKW02, HKC07, HKC08, HCZ03, HCZ06, and HCZ08) or phylogroup II (9 specimens of HKS05, HKB07, HKW03-HKW5, HKC04-HKC06, and HKC09), whereas allele HA 3 was composed of two specimens of phylogroup II (HCZ04 and HCZ07), and allele HA 4 consisted of two specimens of phylogroup II (HCD09 and HCD10).

Thus, we concluded that faster evolutionary rate of mtDNA genome and stronger influence by bottleneck effects on mtDNA haplotypes have resulted in the two, sympatric phylogroups of the Chinese water deer in Korea, which were recognized by cytochrome *b* and control region sequences but were not distinguishable by *IRBP* sequences.

Siberian Roe Deer

From the comparison of control region sequences of the Siberian roe deer from Korean Jeju Island, far-eastern Russian Amur, and western Siberian Kurgan, Koh and Randi (2001) found the genetic distinctness of the Jeju roe deer, with the average distance of 1.51% (range from 0.74% to 2.24%). From this study based on *IRBP* sequences from the Siberian roe deer in East Asia (Figure 3), Jeju island roe deer, with monogenic allele of CA 2, was distinct from the Siberian roe deer from the Korean peninsula, northeastern China, far-eastern Russian Vladivostok, and Mongolia, with monogenic allele of CA 1. The sequences of four variable sites (no. 563, 626, 690, and 909) of the two alleles (CA 1 and CA 2) were CCGG and TTAA, respectively, indicating that Jeju Island roe deer is distinct at four fixed sites, and the maximum Tamura-Nei nucleotide distance was 0.33%.

The Korean Jeju Island, with the Halla Mountain (1,950 m above sea level) in the central part, originated by a series of volcanic activities at the end of the Tertiary, about two million years ago (Park, 1985), although the Yellow-Sea, separating Jeju Island and the Korean peninsula, was the land during the most recent glacier, about 20,000-10,000 years ago (Chung, 2007), indicating that the Jeju Island was connected several times to the mainland during the Pleistocene and that it was recently separated again about 10,000 years ago. The evolutionary rates of nuclear genes are slower than those of mitochondrial DNA (mtDNA) sequences (Steppan et al., 2005), and the rate of nucleotide substitution in mammalian cytochrome *b* gene is 2.1% per million years (Irwin et al., 1991), whereas control region is the most divergent regions among mtDNA sites (Lopez et al., 1997).

The Jeju island roe deer was distinct from the Siberian roe deer of other regions, with four fixed site differences and the maximum nucleotide distance of 0.33% in *IRBP* sequences (Figure 3) and 1.51% in control region sequences (Koh and Randi, 2001). In addition, the Siberian roe deer in Korea and Jeju Island prefer the montane area of middle and higher altitudes. Therefore, the most plausible explanation for genetic distinctness of the Jeju Island roe deer is that the ecological barrier, i.e., the Yellow Sea during glacier phase and the sub-montane region of lower altitudes in the foot of the Halla Mountain during inter-glacier phase have isolated Jeju Island roe deer, once the first small population of Siberian roe deer from the Korean peninsula dispersed into the forest region of Jeju Island during the Pleistocene.

Barclay (1935) described the difference of the pelage color of the Siberian roe deer from Korea and classified it as *C. p. ochracea*, whereas Danilkin (1996) classified the Siberian roe deer from the Tien Shan and far-eastern Asia, including Korea, as *C. p. tianschanicus*, without examination of the Siberian roe deer from the Korean peninsula and Korean Jeju Island. In this *IRBP* sequence analysis with the Siberian roe deer from East Asia (Figure 3), we found that the Siberian roe deer from the Korean peninsula, northeastern China, far-eastern Russia, and Mongolia are similar with one another, indicating that these sequencing results support the classification of Korean peninsula roe deer by Danilkin (1996) and not by Barclay (1935).

On the other hand, we confirmed from our *IRBP* and control region sequence analyses of the Siberian roe deer from Asia that the Jeju Island roe deer is an allopatric phylogroup within *C. p ochracea*, as mentioned above. A subspecies is an aggregate of phenotypically similar populations of a species differing taxonomically from other populations of that species (Mayr and Ashlock, 1991), and it is necessary to perform morphological analysis, including pelage color, to clarify the subspecific status of both the Korean peninsula roe deer and Jeju Island roe deer.

CONSERVATION OF TWO ENDEMIC SUBSPECIES OF DEER IN KOREA

Avise and Walker (1999) reported that about 140 of 252 species of vertebrates were subdivided clearly into two or more highly distinctive matrilineal phylogroups with at least 0.6% sequence diversity in mtDNA and that the vast majority of these phylogroups were localized geographically. Bradley and Baker (2001) concluded that a genetic distance of less than 2%, based on cytochrome *b* gene, was typical of populational and intraspecific variation. In addition, the lineage sorting from polyphyly through paraphyly to monophyly must occur before alleles or haplotypes become population-specific (Freeland, 2005).

In our study on the Chinese water deer, the endemic subspecies of *H. i. argyropus* from Korea was composed of two sympatric phylogroups based on cytochrome *b* gene and control region sequences (Koh et al., 2009), indicating that two phylogroups are monophyletic in its mtDNA sequences. But, from *IRBP* sequence analysis, the specimens belonging to the two phylogroups (I and II) were randomly sorted into the four alleles (see Figure 2), indicating that the Chinese water deer from Korea is paraphyletic in their nuclear *IRBP* sequences. On the other hand, Jeju Island roe deer (*C. p. ochracea*) is an allopatric phylogroup, which is distinct in nuclear *IRBP* and mtDNA control region sequences from other Siberian roe deer, including the Korean peninsula, northeastern China, far-eastern Russia, Mongolia, and western Siberian Kurgan [Figure 3 and Koh and Randi (2001)], indicating that Jeju Island roe deer is monophyletic in its nuclear and mtDNA sequences.

Moritz (1994) designated a population that is reciprocally monophyletic for mtDNA haplotypes and shows significant divergence of allele frequencies at nuclear loci as ESU. Pennock and Dimmick (1997) noted that protection under the U. S. Endangered Species Act is extended to "any subspecies of fish or wildlife or plants, and any distinct population segment of any species of vertebrate fish or wildlife which interbreeds when mature." Thus, we concluded that two endemic subspecies of Korean deer, with genetic distinctness, are ESUs, which need special protection for their conservation. In addition, two Korean deer species are unique in their distributional aspects. Korean water deer is one of two natural populations of the Chinese water deer in the world and is geographically isolated population. Jeju Island roe deer is a peripheral population and is ecologically and geographically isolated population.

However, the protection status of two endemic subspecies of deer from Korea, showing distinctness in DNA sequences and distributional aspect, is in striking contrast. The Chinese water deer from the Korean peninsula (*H. i. argyropus*), which is abundant in South Korea, is hunted as one of a pest for crops and a game mammal (Korean Ministry of Environment, 2005), although the other natural population in China is listed as near threatened in the IUCN Red List (Grubb, 2005). On the other hand, Jeju Island roe deer, which was rare and recently became abundant, is protected by the local government of Jeju Province, although the Siberian roe deer (*C. pygargus*) distribute widely in Eurasia (IUCN category of the least concern).

Gardenfors (2001) showed the proposed conceptual scheme of the procedures for assigning an IUCN Red List category at a national or regional level and noted that protection categories can be downgraded or upgraded depending on the criterion of contact between local population and other infra-specific populations. Thus, in order to determine protection tactics for two subspecies of deer in South Korea and other deer species from other regions as well, we recommend the adoption of another criterion to upgrade into threatened category, i.e., the evaluation of distributional and evolutionary uniqueness of global scale, when the local population size is large (> 1,000) enough to fit the IUCN guideline for the least concern category.

REFERENCES

Avise JC, Walker D (1999) Species realities and numbers in sexual vertebrates: perspectives from an asexual transmitted genome. *Proc. Natl. Acad. Sci. USA* 96: 992-995.

Barclay EN (1935) The roe deer of Korea. *Ann. Mag. Nat. Hist.* (London) 15: 626-627.

Bradley RD, Baker R J (2001) A test of the genetic species concept: cytochrome *b* sequences and mammals. *J. Mamm.* 82:960-973.

Chapman RW (1989) Mitochondrial and nuclear gene dynamics of introduced populations of *Lepomis macrochirus*. *Genetics* 123:399-404.

Chappell DE, van den Bussche RA, Krizan J, Patterson B (2004) Contrasting levels of genetic differentiation among populations of wolverlines (*Gulo gulo*) from northern Canada revealed by nuclear and mitochondrial loci. *Conservation Genet.* 5:759-767.

Chung CH (2007) Vegetation responses to climate change on Jeju island, South Korea, during the last deglaciation based on pollen record. *Geosci. J.* 11:147-155.

Corbet GB (1978) The mammals of the Palaearctic region: a taxonomic review. *British Mus (Nat. Hist.),* Cornell University Press, London, 314pp.

Danilkin A (1996) *Behavioral ecology of Siberian and European roe deer.* Chapman and Hall, London, 277pp.

Frankham R, Ballou JD, Briscoe DA (2004) *A primer of conservation genetics.* Cambridge University Press, Cambridge, 220pp.

Freeland JR (2005) *Molecular ecology.* John Wiley and Sons, Ltd., West Sussex, England, 388pp.

Gardenfors U (2001) Classifying threatened species at national versus global levels. *TREE* 16:511-516.

Grubb P (2005) Order Artiodactyla. In: Wilson DE, Reeder DM (eds) Mammal species of the world. *A taxonomic and geographic reference*, 3rd ed. The Johns Hopkins University Press, Baltimore, pp 654-655.

Irwin DM, Kocher TD, Wilson AC (1991) Evolution of the cytochrome *b* gene of mammals. *J. Mol. Evol.* 32:128-144.

Koh HS, Randi E (2001) Genetic distinction of roe deer (*Capreolus pygargus* Pallas) sampled in Korea. *Mamm. Biol.* 66: 371-375.

Koh HS, Lee BK, Wang J, Heo SW, Jang KH (2009) Two sympatric phylogroups of the Chinese water deer (*Hydropotes inermis*) identified by mitochondrial DNA control region and cytochrome *b* gene analyses. *Biochem. Genet.* 47: 860-867.

Koh HS, Oh JG, Lee BG, Jang KH, In ST, Kim GC (2010) Species identification of the sika deer (*Cervus nippon* Temminck 1838) in deer farms from northeast China, North Korea, and Korea, revealed by the sequencing of mitochondrial DNA cytochrome *b* gene. *Bulletin of Natural Sciences* 24:7-11.

Korean Ministry of Environment (2005) *Atlas of rare and endangered wild plants and animals.* Ministry of Environment, Seoul, Korea, 247pp (in Korean).

Lopez VJ, Culver M, Stephens JC, Johnson WE, O`Brien SJ (1997) Rate of nuclear and cytoplasmic mitochondrial DNA sequence divergence in mammals. *Mol. Biol. Evol.* 14: 277-286.

Mayr E, Ashlock PD (1991) *Principles of systematic zoology.* McGraw Hill Inc., New York, 475pp.

Moritz, C (1994) Defining "evolutionary significant units" for conservation. *TREE* 9:373-375.

Park DW (1985) Geology and coasts and mountains in Cheju-island. *Cheju-Do study* (Cheju Province, Korea) II: 321-322 (in Korean).

Pennock DS, Dimmick WW (1997) Critique of the evolutionary significant unit as a definition for "distinct population segments" under the U.S. endangered species act. *Conserv. Biol.* 11:611-619.

Primack RB (1995) *A primer of conservation biology*. Sinauer Associates Inc., Mass, 277pp.

Stanhope MJ, Czelusniak J, Si J, Nickerson J, Goodman M (1992) A molecular perspective on mammalian evolution from the gene encoding interphotoreceptor retinol binding protein, with convincing evidence for bat monophyly. *Mol. Phylogenet. Evol.* 1:148-160.

Steppan SJ, Adkins RM, Spinks PQ, Hale C (2005) Multigene phylogeny of the Old world mice, Murinae, reveals distinct geographic lineages and the declining utility of mitochondrial genes compared to nuclear genes. *Mol. Phylogenet. Evol.* 37:370-388.

Tamura K, Nei M (1993) Estimation of the number of nucleotide substitutions in the control region of mitochondrial DNA in humans and chimpanzees. *Mol. Biol. Evol.* 10: 512-526.

Woon PH (1967) Illustrated encyclopedia of fauna and flora: vol 7, *Mammals*. Ministry of Education, Seoul, 659pp (in Korean).

In: Deer: Habitat, Behavior and Conservation ISBN 978-1-62100-676-3
Editors: A. A. Cahler and J. P. Marsten

Chapter 6

EFFECTS OF SIKA DEER ON VEGETATION IN JAPAN WITH REFERENCE TO DIFFERENCES BETWEEN NORTHERN AND SOUTHERN JAPAN

Seiki Takatsuki*
Laboratory of Wildlife Ecology and Conservation,
School of Veterinary and Medicine, Azabu University,
Fuchinobe, Chuo-ku, Sagamihara, Kanagawa, Japan

ABSTRACT

The Japanese Archipelago is elongated in a north–south direction and contains various vegetation zones. Sika deer (*Cervus nippon*) range in these zones. In northern Japan, deciduous oak forest and beech forest prevail. The floors of these forests are densely covered by dwarf bamboos, which, because they are evergreen, provide important forage for sika deer during winter. The removal of dwarf bamboos by deer sometimes favors seedlings and facilitates forest regeneration. Northern deer undertake vertical migrations according to snow during winter. Winter ranges are therefore highly populated, resulting in the barking of trees. In southwestern Japan, lowlands are covered by evergreen oak forests with dark and poorly vegetated floors. However, the existing

* Email: takatuki@azabu-u.ac.jp, Phone: +81-42-850-2451, fax: +81-42-850-2487.

understory plants are evergreen, and deer food resources show less seasonal fluctuation than in northern Japan. Southern deer are sedentary and do not show seasonal changes in density. Mountains in southern Japan are covered by evergreen trees, and *Sasamorpha borealis*, a dwarf bamboo, grows on the forest floors. When deer densities are high, as is often the case on small islands, many plants are decimated by deer grazing and are replaced by unpalatable plants or by *Zoysia japonica*, a lawn sward. Recently, sika deer have expanded their range into snowy areas and alpine zones.

Keywords: dwarf bamboo, forest regeneration, grazing, *Sasa nipponica*, sika deer, unpalatable plants, *Zoysia japonica*

INTRODUCTION

The Japanese archipelago is elongated in a north–south direction, therefore extending from the warm temperate zone dominated by evergreen oak forests in the southwest through the cool temperate zone dominated by deciduous forests of oaks and beech in the north, with coniferous forests in higher areas and in Hokkaido (the northern island). Sika deer (*Cervus nippon*) inhabit almost all types of vegetation. The deer population decreased up to the 1970s, particularly in the northern part of Honshu (the main island) due to over-hunting during the snow season, and a wide area of unpopulated habitat resulted. Recently, however, populations have been increasing and ranges expanding due to the reduction in snow cover caused by global warming and a decline in the number of hunters. The sika deer range expanded by 30% from the 1970s to the 2000s (Japan Wildlife Research Center, 1980; Biodiversity Center of Japan, 2004). Accordingly, damage to agriculture and forestry as well as to natural vegetation has become serious. This paper reviews the recent effects of deer grazing on vegetation of the Japanese archipelago.

I. NORTHERN JAPAN

Mountains in central and northern Japan are dominated by deciduous broad-leaved forests including oaks and beech, and lower hills are covered by coppiced forests of oaks. The latter are also covered by planted conifers. These deciduous forests often include ground cover in the form of dwarf bamboos.

Figure 1. *Sasa nipponica*. Dwarf bamboos of the genus *Sasa* have reduced numbers of large leaves.

Dwarf bamboos grow widely in the forests of East Asia, but those in Japan grow particularly densely. The leaves of Japanese dwarf bamboos are large, and there are far fewer leaves per culm compared with dwarf bamboos in China and Taiwan. This is particularly true of the genus *Sasa* (Figure 1). It is unique in that evergreen grasses grow densely in deciduous broad-leaved forests. Due to the dense cover of dwarf bamboos, the forest floors are shaded, and the growth of other shrubs and herbaceous plants is suppressed.

The genus *Sasa* includes three sections (Suzuki, 1961). Dwarf bamboos in the section *Macrochlamys*, which grow in snowy areas, are tallest, reaching as high as 3 m, and are densely branched. The leaves live for several years. In zones of intermediate snow depth, dwarf bamboos of the section *Sasa* grow. Dwarf bamboos in this section are 1–2 m high, and the branches are not so dense. On the Pacific side of Japan, where snow depth is less than 50 cm, dwarf bamboos in the section *Crassinodi*, including *Sasa nipponica*, grow. These are shorter than 70 cm and do not branch. Until recently, sika deer ranged within this "*Sasa nipponica* zone" (Takatsuki, 1992a, 2009c). Since the legs of sika deer are about 60 cm long and the hooves are small, they cannot move in deep snow (Takatsuki, 1992b). Hunting such deer is easy, and the deer population is vulnerable. Consequently, the distributions of sika deer and of *Sasa nipponica*, both of which are adapted to a thin-snow habitat, principally correspond.

The lack of branches in *Sasa nipponica* has been explained as an adaptation to the dry and cold winter environment of the Pacific side of northern Japan (Suzuki, 1961). The turnover rate of above-ground parts is rapid, and it grows like a biennial herb. Although sika deer do not always prefer this dwarf bamboo during summer when other plants are abundant, they are dependent on it during winter (Takatsuki, 1983, 1986, Campos-Arceiz and Takatsuki, 2005). According to its rapid turnover rate, *Sasa nipponica* is tolerant to defoliation and is able to regrow (Takatsuki, 2009c). This enables its "sustainable use" by sika deer. *Sasa nipponica* is tolerant to deer grazing because grazing removes other co-growing shrubs and trees that shade *Sasa nipponica* and suppress its growth when no grazing occurs (Takatsuki 2006, Figure 2). Due to these facts, the area of northern Japan where *Sasa nipponica* and sika deer habitats overlap is regarded as a "Sika deer–*Sasa nipponica* biome". The ecology of sika deer between northern and southern Japan was compared by Takatsuki (2009c).

Figure 2. Inside a deer-proof fence, *Sasa nipponica* initially recovers, but after several years, other shrubs colonize and suppress its growth.

In fact, *Sasa nipponica* is an important food in many places including the Shiranuka Hills in Hokkaido (Campos-Arceiz and Takatsuki 2005), Mt. Goyo in northern Honshu (Takatsuki 1986), Nikko (Takatsuki, 1983), and Tochigi Prefecture (Takatsuki and Ueda, 2007) and Ohdaigahara in Wakayama, western Japan (Yokoyama and Shibata, 1997). It occupies more than 70–80% of the montane zone in winter and spring but accounts for less of the percentage cover in subalpine zones and lower hilly areas.

Sasa palmata, a representative dwarf bamboo of the section *Sasa*, exhibits slower turnover of its aboveground parts, and therefore suffers more harm from grazing or defoliation damage (Matsui, 1963). Nomiya et al. (2002) reported a marked decrease in *Sasa palmata* at a riparian forest in central Japan caused by deer grazing. Because dwarf bamboos have great effects on the forest ecosystems of Japan, their reduction by deer grazing would affect many other organisms and their interrelationships.

Snow and Sika Deer

In northern Japan, many areas are snowy, and overwintering is important for sika deer. Recent global warming, however, seems to have reduced winter mortality of deer, particularly of fawns, resulting in expansion of the range of sika deer into snowier areas (Biodiversity Center of Japan 2004).

Snow forces sika deer to leave foraging grounds during winter, and sika deer show vertical migrations in snowy areas (Yabe and Takatsuki, 2009). The distances of movement are usually less than several kilometers. Deer densities are quite high in the winter ranges, and the deer often starve in such places and feed even on bark (Figure 3). It is characteristic in snowy areas for deer to range widely at low densities during summer. The deer populations in such areas are in good nutritional condition, and pregnancy rates are high.

Figure 3. A trunk of *Abies homolepis* debarked by sika deer in Oku-Nikko, central Japan.

Effects on Vegetation

Several phases of vegetation change are recognized according to deer density. At densities lower than 1 deer/km^2, only slight effects are found, and some trees and shrubs are partly deformed. Negative effects on plant growth are not observed. At densities of about 10 deer/km^2, woody plants show pruned "bonsai-like" forms, and feeding prints are frequently found. Seedlings are removed, and forest regeneration is suppressed. For example, the seedlings of conifers such as *Abies homolepis* and *Picea jezoensis* var. *hondoensis*

(Kumar and Shibata, 2007, Ito and Hino 2004) and broad-leaved trees (Tamura 2008) are reduced by deer browsing. Above this density, preferred plants such as *Aucuba japonica* (Aucubaceae) disappear, and the species composition of some plant communities is reduced. Alternatively, unpalatable plants become more apparent. When dwarf bamboos are removed by heavy deer grazing, some tree seedlings and saplings are favored by the improved light conditions and exhibit improved survival (Ito and Hino, 2005).

Since summer in Japan is hot and humid, plants grow well. The mean monthly temperatures in July, August, and September in Tokyo, for example, are 29.0, 30.8, and 26.8°C, and the monthly mean rainfalls are 161.5, 208.5, and 163.1 mm, respectively (annual mean 16.1°C, 1466.8 mm). These conditions promote vigorous plant growth, and therefore soil erosion and landslides do not usually occur. They do, however, occur in locations with steep topography, where deer density is not always high.

Recently, sika deer ranges have expanded and invaded subalpine forests where they cause damage to trees by barking (Yokoyama et al., 2001).

It is uncommon for deer densities to exceed 50 deer/km^2 on the main islands, but densities can be higher on small islands. In such cases, forest regeneration is suppressed by deer browsing (Takatsuki and Gorai, 1994). Palatable plants are reduced, and unpalatable plants and *Zoysia japonica*, a lawn-type community, replace them. Unpalatable plants include:

Toxic plants : *Senecio cannabifolius* (Asteraceae), *Ligularia dentate* (Asteraceae), *Primula japonica* (Primulaceae), *Chloranthus serratus* (Chloranthaceae), *Chloranthus japonicus* (Chloranthaceae), *Scopolia japonica (*Solanaceae*)*, *Cynanchum caudatum* (Asclepiadoideae), *Paeonia obovata (*Ranunculaceae*)*, *Boenninghausenia japonica* (Rutaceae), *Phytolacca esculenta* (Phytolaccaceae), *Pieris japonica* subsp. *japonica* (Ericaceae), *Leucothoe grayana* (Ericaceae), *Rhododendron molle* subsp. *japonicum* (Ericaceae), *Pachysandra terminalis* (Buxaceae), *Pterostyrax hispida* (Styracaceae); *Arisaema limbatum* var. *ionostemma* (Arales), *Arisaema urashima* (Arales), *Veratrum album* subsp. *oxysepalum* (Liliaceae); *Pteridium aquilinum* (Dennstaedtiaceae), and *Hypolepis punct*ata (Dennstaedtiaceae).

Odorous plants: *Perilla frutescens* var. *citriodora* (Lamiaceae) and *Caryopteris divaricata* (Verbenaceae).

Spiny plants: *Cirsium amplexifolium* var. *muraii* (Asteraceae), *Zanthoxylum piperitum* (Rutaceae), *Berberis thunbergii* (Berberidaceae), and *Aralia elata* (Araliaceae).

The *Zoysia japonica* Community

In open places inhabited by sika deer at high densities (>100 deer/km^2), tall grasses are replaced by lawns of the *Zoysia japonica* community (Figure 4). *Zoysia japonica* is able to reproduce and is tolerant to grazing. However, it is intolerant to shading (Hirayoshi et al., 1969). Therefore, deer grazing favors *Zoysia japonica* by removing competitive plants that could potentially grow tall and shade low-growing plants. *Zoysia japonica* produces comparatively large seeds upon which sika deer preferentially feed, and many seeds are recovered in deer pellets. A feeding experiment showed that about 40% of the seeds ingested were germinable. Seeds with their fruit coats partly removed had higher germinability (Hirayoshi et al., 1969). It is thus plausible that sika deer function to disperse *Zoysia* seeds. This seems to be a good example of the "foliage is a fruit" hypothesis (Janzen, 1984). In fact, the *Zoysia* community has expanded in range at locations with high densities of deer on Kinkazan Island (Ito and Takatsuki, 2009). Short-growing, sun-loving plants grow in *Zoysia* lawns, including *Hydrocotyle ramiflora* (Apiaceae), *Carex leucochlora(Cyperaceae)*, *Moehringia lateriflora* (Caryophyllaceae), *Stellaria alsine* var. *undulata* (Caryophyllaceae), *Viola obtusa* (Violaceae), *Haloragis micrantha* (Haloragaceae), *Swertia japonica* (Gentianaceae), *Plantago asiatica* (Plantaginaceae), and *Plantago camtschatica* (Plantaginaceae).

Figure 4. View of the *Zoysia japonica* community developing under heavy grazing conditions.

New Phases

In the last three decades, sika deer ranges have expanded into snowy areas where they did not previously occur. Oze mire in mid-Japan is known as the biggest mire in Japan and as one of the country's snowiest places. Therefore, it was surprising that sika deer were first found there in the late 1990s (Naito and Kimura, 1998). The number of deer has been increasing (Li et al., 2001). The effects of the deer on the mire have become marked since 2000, and the deer have been found to dig the mire to take the rhizomes of *Menyanthes trifoliate* (Menyanthaceae, Igarashi, and Takatsuki, 2008, Figs. 5, 6). Because this plant grows in deep ponds in the mire, the digging of the rhizomes changes the water flow in the mire and affects the growth of other mire plants. This seems to be an example of "ecosystem engineering" *sensu* Jones et al. (1994).

Figure 5. A female sika deer digging a mire in the Oze Area. Photo taken by automatic camera (T. Igarashi).

Figure 6. A mire of Oze National Park dug by sika deer.

In the last two decades, sika deer have invaded alpine zones above 2000 m. Alpine meadows and woody plants like *Betula ermanii* and *Sorbus sambucifolia* are reduced and replaced by unpalatable plants or grasses like *Calamagrostis hakonensis* (Poaceae, Figure 7)), which are tolerant to grazing (Chubu Forest Management Bureau, 2007).

Figure 7. A deer-proof fence near Hijiridaira hut of South Alps of Japan. Many unpalatable forbs recover inside the fence. Grasses including *Calamagrostis hakonensis*, a grazing-tolerant grass, are dominant outside the fence. Photo by Y. Kagamiuchi.

Figure 8. A slope between Mt. Irou and Mt. Chausu of South Alps of Japan dominated by unpalatable forb *Veraturum album* subsp. *oxysepalum*. This slope was once vegetated by various palatable forbs before deer increase. Photo by Y. Kagamiuchi.

Plants decreasing in alpine meadows include *Trollius japonicus* (Ranunculaceae), *Anemone narcissiflora* (Ranunculaceae), *Ranunculus acris* var. *nipponicus* (Ranunculaceae), *Scabiosa japonica* var. *alpina* (Dipsacaceae), *Epilobium angustifolium* (Onagraceae), *Bistorta vipipara* (Poligonaceae), *Geranium yesoemse* var. *nipponicum* (Geraniaceae), *Angelica matsumurae* (Apiaceae), *Pedicularis chamissonis* var. *japonica* (Scrophulariaceae), *Pedicularis resupinata* subsp. *teucriifolia* (Scrophulariaceae), *Veronicastrum japonicum* (Scrophulariaceae), *Gentiana makinoi* (Gentianaceae), *Lychnis gracillima* (Caryophyllaceae), *Dactylorhiza aristata* (Orchidaceae), and *Fritillaria camtschatcensis* (Liliaceae).

Unpalatable plants increasing in alpine meadows include *Farfugium japonicum* (Asteraceae), *Tephroseris takedana* (Asteraceae), *Senecio nemorensis* (Asteraceae), *Cirsium senjoense* (Asteraceae), *Aconitum micranthum* (Ranunculaceae), *Veratrum album* subsp. *oxysepalum* (Liliaceae, Figure 8), *Dryopteris expansa* (Dryopteridaceae), and *Coniogramme intermedia* (Adiantaceae).

Recovery of palatable forbs was found in deer-proof fences (Figure 7).

II. SOUTHEASTERN JAPAN

The climax vegetation in southern Japan is evergreen oak forests. These forests are dark, and forage plants for sika deer in the understory layer are poor. Secondary forests regrown from sprouts after logging are deciduous forests dominated by oak trees and pine forests in dryer habitats. The forest floors are richer in these secondary forests than in evergreen forests. Mountain areas are dominated by deciduous forests whose floors are often covered by *Sasamorpha borealis*, a dwarf bamboo. These areas are less snowy than northern Japan, which affects the ecology of sika deer.

The forest floors are dark, and therefore plant biomass is limited even in summer, but is not greatly reduced in winter. Evergreen plants like *Ophiopogon japonicus* (Liliaceae) and *Eurya japonica* (Theaceae) grow, and winter food is fairly abundant. Thus, seasonal fluctuations in carrying capacity are small, and forage supply is more stable than in northern Japan. Seasonal vertical migrations are less marked than in northern Japan (Yabe and Takatsuki, 2009).

Ecology of Southern Sika Deer

The home ranges of southern sika deer are smaller in size, and the distances between summer and winter ranges are less than those in northern Japan (Yabe and Takatsuki, 2009). Food supply is stable, and sika deer can eat evergreen plants during winter (Ikeda et al., 2001, Ikeda 2002). Accordingly, damage to agriculture and forestry occurs widely and lightly, making deer management difficult. Nevertheless, the deer density sometimes becomes high. In Kyushu Mountains in Kyushu (the southern island), for example, *Sasamorpha borealis* was greatly reduced by recent heavy grazing by the deer (Murata et al., 2009). Although this represents a type of forest damage, the situation is more complex than it initially appears, as the reduction in dwarf bamboos can promote the growth of tree seedlings (Murata et al., 2009), which may foster forest regeneration.

Islands Inhabited by Sika Deer

In southeastern Japan, several small islands are inhabited by sika deer at high densities. The following unpalatable plants replace others as dominant in some plant communities.

Toxic plants: *Crassocephalum crepidioides* (Asteraceae), Saururus chinensis (Saururaceae), *Arisaema ringens* (Araceae); *Maesa japonica* (Myrsinaceae), *Triadica sebifera* (Euphorbiaceae); *Arachniodes aristata* (Dryopteridaceae).

Odorous plants: *Lysimachia sikokiana* (Primulaceae); *Neolitsea aciculate* (Laureaceae) and *Neolitsea sericea* (Laureaceae).

Spiny plants: *Rubus microphyllus* (Rosaceae), *Rubus sieboldii* (Rosaceae), and *Aralia elata* (Araliaceae).

Hard serration: *Carex lenta* (Cyperaceae).

The *Zoysia japonica* community develops in open habitats where other low-growing plants are favored by the heavy grazing of sika deer in the *Zoysia* community, including *Centella asiatica* (Apiaceae), *Dichondra repens* (Convolvulaceae), *Sagina maxima* (Caryophyllaceae), and *Plantago asiatica* (Plantaginaceae).

Conclusion

Effects of sika deer on the vegetation of the Japanese archipelago were reviewed. It was shown that situations vary considerably between northern and southeastern Japan. In order to monitor the effects of deer and to conserve plants and plant communities, deer-proof fences have been established. In several places, the plants inside and outside the fences differ greatly. In some places, rare plants that had been regarded as "extinct" have been "restored" inside fences (Tamura, 2005). Twelve extinct or endangered plants appeared 4 years after fencing in the Tanzawa Mountains, central Japan. This simply means that the plants were not extinct but survived, probably in dormant forms such as rhizomes or seeds. If heavy grazing by deer continues over long periods, true extinctions can occur. Population increases of sika deer continue, while human populations in rural and mountain areas, as well as hunter populations, are rapidly decreasing in Japan. The deer-vegetation relationship is facing a difficult phase.

Acknowledgments

Mr. T. Igarashi and Mr. Y. Kagamiuchi permitted to use the photos. This research was partially supported by a research project grant awarded by the Azabu University.

References

Biodiversity Center of Japan. (2004) A Basic Survey of Natural Environment Conservation, No. 6. *Report of Distributions of Mammals*. Ministry of Environment, Japan. (in Japanese).

Campos-Arceiz, A. and Takatsuki, S. (2005). Food habits of sika deer in the Shiranuka Hills, eastern Hokkaido - a northern example among the north-south variations of food habits in sika deer -. *Ecological Research*, 20, 129-133.

Chubu Forest Management Bureau. (2007). *Report of sika deer damages in the reserve forest of South Alps of Japan*. (in Japanese).

Hirayoshi, I., Matsumura, M. and Iwata, E. (1969). Fundamental studies on artificial propagation by seeding of useful wild grasses in Japan, 1. Some experiments on germinability of Japanese lawn grass (*Zoysia japonica* Steud.). *Research Report of Department of Agriculture of Gifu University*, 28, 239-251. (in Japanese with English summary).

Igarashi, T. and Takatsuki, S. (2008). Effects of defoliation and digging caused by sika deer on the Oze mires of central Japan. *Biosphere Conservation*, 9, 9-16.

Ikeda, K. (2002). The food habits of sika deer in a coniferous plantation area, northern Kyushu. *Journal of Japanese Forestry Society*, 84, 175-179.

Ikeda, K., Koizumi, T., Yabe, T., Miyajima, J., Sanui, T., Yoshioka, S., Yoshimoto, K., Sumiyoshi, H., and Tajitsu, H. (2001). Ecology and damage control of sika deer in Kyushu. *Forest Pests*, 50, 1-19. (in Japanese).

Ito, H. and Hino, T. (2004). Effects of deer, mice and dwarf bamboo on the emergence, survival and growth of Abies homolepis (Piceaceae) seedlings. *Ecological Research*, 19, 217-223.

Ito, H. and Hino, T. (2005). How do deer affect tree seedlings on a dwarf bamboo-dominated forest floor? *Ecological Research*, 20, 121-128.

Ito, T. Y. , M. Shimoda, and S. Takatsuki. 2009. Productivity and foraging efficiency of the short-grass (*Zoysia japonica*) community for sika deer. In D. R. McCullough, Takatsuki, S. and Kaji, K. (Eds.), *Sika Deer: Biology and Management of Native and Introduced Populations* (pp. 145-157). Tokyo: Springer.

Janzen, D. H. (1984) Dispersal of small seeds by big herbivores: foliage is the fruit. *American Naturalist*, 123, 338-353.

Japan Wildlife Center. (1980). A Basic Survey of Natural Environment Conservation, No. 2. *Report of Distributions of Animals (Mammals).* Ministry of Environment, Japan. (in Japanese).

Jones, C. G., Lawton, J. H. and Shachak, M. (1994). Organisms as ecosystem engineers. *Oikos*, 69, 373-386.

Kumar, S. and Shibata, E. (2007). Establishment and growth of coniferous seedlings in an altered forest floor after long-term exclusion of deer. *Journal of Forest Research*, 12, 306-311.

Li, Y., Maruyama, N., and Koganezawa, M. (2001). Factors explaining the extension of the sika deer's range in Nikko, Japan. *Biosphere Conservation,* 3, 55-69.

Matsui,Y. (1963). Management of dwarf bamboos for forestry and use for them as resources in Hokkaido. *Annual Report of Hokkaido Branch, Forestry Station of Ministry of Agricultural and Forestry, Japan.* (in Japanese).

Murata, I., Sakurai, S., Kubota, K., Inoue, S., Tashiro, N., Enoki, T., Utsumi, Y. and Inoue, S. (2009). Effects of sika deer (*Cervus nippon*) and dwarf bamboo (*Sasamorpha borealis*) on seedling emergence and survival in cool-temperate mixed forests in the Kyushu Mountains. *Journal of Forest Research,* 14, 296-301.

Naito, T. and Kimura, Y. (1998). Sika deer in the Oze area. In Research Commission of Academic and Comprehensive Study of the Oze Area (Ed.), *A Comprehensive Study of the Oze Area*, (pp. 725-739). Maebashi, Oze Research Commission Group. (in Japanese).

Nomiya, H., Suzuki, W., Kanazashi, T., Shibata, M., Tanaka, H. and Nakashizuka, T. (2002). The response of forest floor vegetation and tree regeneration to deer exclusion and disturbance in a riparian deciduous forest, central Japan. *Plant Ecology,* 164, 263-276.

Suzuki, S. (1961) Ecology of Bambusaceous genera *Sasa* and *Sasamorpha* in the Kanto and Tohoku districts of Japan, with special reference to their geographical distribution. *Ecologocal Revew*, 15, 131-147.

Takatsuki, S. (1983). The importance of Sasa nipponica as a forage for Sika deer (*Cervus nippon*) in Omote-Nikko. *Japanese Journal of Ecology*, 33, 17-25.

Takatsuki, S. (1986). Food habits of Sika deer on Mt. Goyo. *Ecological Research,* 1, 119-128.

Takatsuki, S. (1992a) *A Herd of Sika Deer Living in the North.* Tokyo: Doubutsu-sha Publ. Inc., Tokyo. (in Japanese).

Takatsuki, S. (1992b). Foot morphology and distribution of Sika deer in relation to snow depth. *Ecological Research*, 7, 19-23.

Takatsuki, S. (2003). Use of mires and food habits of sika deer in the Oze Area, central Japan. *Ecological Research,* 18, 331-338.

Takatsuki, S. (2006) *Ecological History of Deer.* Tokyo University Press. (in Japanese)

Takatsuki, S. 2009a. A 20-year history of sika deer management in the Mt. Goyo area northern Honshu. In D. R. McCullough, Takatsuki, S. and Kaji, K. (Eds.), *Sika Deer: Biology and Management of Native and Introduced Populations* (pp. 365-373). Tokyo: Springer.

Takatsuki, S. (2009b). Effects of sika deer on vegetation in Japan: *a review. Biological Conservation*, 142, 1922-1929.

Takatsuki, S. (2009c) North-south variations in sika deer ecology as a forest-dwelling cervid. In D. R. McCullough, Takatsuki, S. and Kaji, K. (Eds.), *Sika Deer: Biology and Management of Native and Introduced Populations* (pp. 217-230). Tokyo: Springer,.

Takatsuki, S. and Gorai, T. (1994). Effects of Sika deer on the regeneration of a *Fagus crenata* forest on Kinkazan Island, northern Japan. *Ecological Research*, 9, 115-120.

Takatsuki, S. and Ueda, H. (2007). Meso-scale variation in winter food composition of sika deer in Tochigi Prefecture, central Japan. *Mammal Study*, 32, 115-120.

Tamura, J. (2005). Protection of endangered plants from sika deer heribivory using deer-prood fences in the Tanzawa Mountains, central Japan. *Japanese Journal of Conservation Ecology*, 10, 11-17. (in Japanese with English abstract)

Tamura, J. (2008). Effects of fencing for 7 years on the regeneration of tree species after dwarf bamboo on the forest floor has been diminished by sika deer browsing. *Journal of Japanese Forestry Society*, 90, 158-165.

Yabe, T. and Takatsuki, S. (2009). Migratory and sedentary behavior patterns of sika deer in Honshu and Kyushu, Japan. In D. R. McCullough, Takatsuki, S. and Kaji, K. (Eds.), *Sika Deer: Biology and Management of Native and Introduced Populations* (pp. 273-283). Tokyo: Springer.

Yokoyama, S. and Shibata, E. (1997) .The effects of sika-deer browsing on the biomass and morphology of a dwarf bamboo, *Sasa nipponica*, in Mt. Ohdaigahara, central Japan. *Forest Ecology and Management*, 103, 49 - 56.

Yokoyama, S., Maeji, I., Ueda, T., Ando, M. and Shibata, E. (2001). Impact of bark stripping by sika deer, *Cervus nippon*, on subalpine coniferous forests in central Japan. *Forest Ecology and Management*, 140, 93-99.

INDEX

E

F

G

H

I

J

K

L

M

N

O

P

Q

R

S

T

U

V

W

Y

Z